AF306598

Valery Ustyantsev

Permeable fault zones and their role

Valery Ustyantsev

Permeable fault zones and their role

Possibility of application of Prigozhin's principles of nonlinear thermodynamics and Curie principle in geology

ScienciaScripts

Imprint

Any brand names and product names mentioned in this book are subject to trademark, brand or patent protection and are trademarks or registered trademarks of their respective holders. The use of brand names, product names, common names, trade names, product descriptions etc. even without a particular marking in this work is in no way to be construed to mean that such names may be regarded as unrestricted in respect of trademark and brand protection legislation and could thus be used by anyone.

Cover image: www.ingimage.com

This book is a translation from the original published under ISBN 978-620-3-04155-2.

Publisher:
Sciencia Scripts
is a trademark of
International Book Market Service Ltd., member of OmniScriptum Publishing Group
17 Meldrum Street, Beau Bassin 71504, Mauritius
Printed at: see last page
ISBN: 978-620-3-36671-6

Table of contents

Introduction

Back in the early 1920s, Vernadsky wrote "about the need to create a 'science of the future', a science that studies 'the energy of our planet'.

V. I. Popov (1938) distinguished 13 gradations of wave pulsations from large to seismic waves and emphasised that "in the development of large and long-lasting wave oscillations the endless series of subordinate, smaller and more frequent oscillations, in which the body of our planet shakes continuously, are integrated according to the rules of a kind of "natural selection".

The German researcher W. Ebeling states that "questions of the formation of structures belong to the fundamental problems of the natural sciences, and the study of the emergence of structures is one of the most important goals of scientific knowledge".

Geologists work with the consequences of the mechanisms that shape the Earth's system and concentrate minerals when prospecting and exploring. In other words, an important task of geology is to establish cause-effect relationships between geological processes and phenomena occurring in the Earth system.

This concept contributes to the further development of the scientific direction initiated by V.I. Popov and V.V. Bogatsky and their students, which was defined by Academician V.I. Smirnov [1978] as the emergence of "a new original concept of the wave nature of deforming stresses".

Pierre Curie's principle of symmetry and dissymmetry

"In 1884 he published an article on the questions of orderliness and repetition underlying the study of crystal symmetry. This article was followed in the same year by a more general treatment of the same problem. Another article on symmetry and repetition was published in 1885. In the same year, he published a very important theoretical work on crystal formation and on capillary constants of different faces.

From this rapid sequence of works one can see how much Pierre Curie was absorbed by crystal physics. His theoretical and experimental studies in this field are grouped around a very general principle, the principle of symmetry, which he gradually established; this principle received its final expression only in articles published between 1893 and 1895.

This is the wording that has since become a classic.

"When certain causes cause certain effects, the elements of symmetry of causes must show up in the effects they cause."

"When a certain asymmetry is found in any phenomenon, the same asymmetry must also be found in the causes which give rise to it.

"Positions the opposite of these are wrong, at least practically; in other words, the effects may have a higher symmetry than the causes that caused them."

The primary importance of these provisions, very perfect in all their simplicity, is that the elements of symmetry in question apply to all physical phenomena without exception. Guided by an in-depth study of symmetry groups which may exist in nature, Pierre Curie showed how to use these provisions as much geometric as physical ones, in order to foresee whether a particular phenomenon is possible or impossible under given conditions. At the beginning of one of his articles he states the following: "I think that in physics we must introduce the notions of symmetry which are habitual for crystallographers". (Marie Curie, Pierre Curie, Moscow, Nauka, 1968, p.21-22).

I. Prigozhin's basic principles of nonlinear thermodynamics

Synergism is a cooperative action:

"Nature is hierarchically structured into several kinds of open non-linear systems of different levels of organisation: into dynamically stable, adaptive, and most complex - evolving systems. The connection between them is made through the chaotic, non-equilibrium state of neighbouring level systems.

Imbalance is a necessary condition for the emergence of a new organisation, a new order, new systems, i.e. development.

When non-linear dynamic systems are combined, the new entity is not equal to the sum of the parts, but forms a system of a different organisation, or a different level of system.

Common to all evolving systems:

- non-equilibrium; spontaneous formation of new microscopic (local) formations;

- changes at the macroscopic (system) level; the emergence of new system properties;

- the stages of self-organisation and the fixation of new qualities of the system.

Evolving systems are always open and exchange energy and matter with the external environment, which is why the processes of local order and self-organisation take place.

In highly non-equilibrium states, systems begin to perceive influences from the outside that they would not perceive in a more equilibrium state.

Under non-equilibrium conditions, the relative independence of the elements of the system, gives way to a corporate behaviour of the elements: near the equilibrium an element interacts only with its neighbours, far from the equilibrium it "sees" the whole system and, the coherence of the behaviour of the elements increases.

In states far from equilibrium, bifurcation mechanisms come into play: - the presence of short-term bifurcation points of transition to one or another relatively long-term mode of the system - an attractor. It is impossible to predict in advance which of the possible attractors the system will occupy.

Synergetics explains the process of self-organisation in complex systems as follows:

The system must be open.

A closed system, according to the laws of thermodynamics, must eventually reach a state of maximum entropy and cease any evolution.

A fundamental principle of self-organisation is the emergence of new order and complexity of systems through fluctuations (random deviations) of the states of their elements and subsystems. Such fluctuations are usually suppressed in all dynamically stable and adaptive systems, due to negative feedbacks that ensure the preservation of the structure and near-equilibrium state of the system. But in more complex open systems, due to the inflow of energy from outside and the intensification of non-equilibrium, fluctuations increase over time, accumulate, cause a collective behaviour effect of elements and subsystems and finally lead to a "loosening" of the previous order and through a relatively short-lived chaotic system state lead either to the destruction of the previous structure or to the appearance of a new order. Since fluctuations are random, the state of the system after a bifurcation is due to the action of a sum of random factors.

Self-organisation, having as its outcome the formation, through a stage of chaos, of a new order or new structures, can occur only in systems of sufficient complexity, possessing a certain number of interacting elements, having some critical coupling parameters and relatively high values of the probabilities of their fluctuations. Otherwise, the effects of synergistic interaction will be insufficient for the emergence of collective behavior of the system elements and thus the emergence of self-organization. Insufficiently complex systems are not capable of spontaneous adaptation, much less of development, and, when receiving excessive amounts of energy from the outside, lose their structure and irreversibly collapse.

Self-organisation in complex systems, transitions from one structure to another, the emergence of new levels of organisation of matter, are accompanied by the breaking of symmetry. When describing evolutionary processes, it is necessary to abandon the symmetry of time characteristic of fully deterministic and reversible processes in classical mechanics.

Self-organization in complex and open - dissipative systems, to which life and mind belong, leads to irreversible destruction of old structures and systems and to emergence of new ones, which, along with the phenomenon of non-decreasing entropy in closed systems, determines the presence of "arrow of time" in Nature". *(Glensdorf P., Prigozhin I.* Thermodynamic theory of structure, stability and fluctuations. Moscow: Mir, 1973; *Nikolis P., Prigozhin I.* Self-organization in non-equilibrium systems; *Prigogine I., Stenger I.* Order out of chaos: A new dialogue between man and nature. - Moscow: Progress, 1986. *Nikolis, P.; Prigogine, I.* Self-organization in nonequilibrium systems. - Moscow: Mir, 1979).

The Earth's crust and the formation of the geochemical system of hydrocarbons
(V.I. Vernadsky, 1934)

"XVIII-XVIII centuries. At that time, the surface crust of the Earth was thought to be the remnant of the first solidification of the liquid molten mass of our planet - in ancient, entirely hypothetical periods of its existence. These walking but erroneous ideas have nothing to do with the real crust of the Earth, which is studied in geochemistry.

It is easy to see that the Earth's crust shows no sign of a primary, hypothetical crust anywhere. It has only one name in common with it. The Earth's crust is the upper region of the planet, complex in its structure and in its origin, empirically ascertainable. From below, it is limited by the isostatic surface, and has a thickness of 60-100km. It can be dissected into separate shells, starting from the top:

1. ionosphere, stratosphere, biosphere;

2. stratisphere, a metamorphic shell.

3. granite shell, basalt (main) shell. These shells define the area of the globe within which only geochemical processes take place. Below lies the Earth's matter, of which we have a very imprecise and incomplete understanding. With the current level of scientific knowledge, it is unlikely that one can proceed confidently in the geochemical investigation based on our dominating concepts regarding the state of matter and the nature of chemical processes that exist and may exist below the Earth's crust. In this field a great scientific work is going on, but now with the use and hope for success we can not do below the crust in the geochemical phenomena. But at the same time by doing so we kill the quest and create the illusion of knowledge where scientific guesswork and extrapolation prevail. The Earth's crust is a self-sustaining entity with a certain degree of organisation, an autocracy - processes that begin and end within it. Their study comes first as the task of the day of geochemistry. Only by taking this approach can we determine how our empirical geochemical material reveals the influence of the deeper layers. A key feature of the Earth's crust is that it is the only region of the planet where the physical states of matter we all know - and which determine life and the environment around it - exist and can manifest:

- solid, liquid and gaseous. This is the only area of the planet where they can all exist. It is the only region of the planet where they can all exist. It is correct to take this feature as a baseline, to define the region of geospheres, as it is possible that there is no coincidence of the crustal region with the boundary of isostatic surface, which is often taken as lower boundary of the Earth crust. Already at 60 km down from the geoid level, under the landmass, pressure reaches about 30 thou. at/cm2 , at which point the distinction between solid (crystalline), liquid, and gaseous states disappears. Plastic and elastic matter has new manifestations, completely alien to us. We cannot realistically even imagine it, because we are not able to express the world figuratively under such pressures. The real boundary showing sharp change of properties of the planet substance, in the light of new data, goes down to 100-120 km and in such case is clearly out of limits of Earth crust, if we put into base of it the coexistence of three phases of Earth substance.

The most important geochemical properties of the geospheres of the Earth's crust:

1. The chemical elements in them are in stable dynamic equilibria, different for each geosphere.

2. They move - really move - in the course of geological time, from one geosphere to another.

3. The state of the elements - the equilibria that correspond to them - change when chemical elements move from one geosphere to another. This change occurs periodically, slowly or suddenly.

The history of most chemical elements in the Earth's crust - all cyclic, much of the dispersed and highly radioactive elements and REE - is characterized by closed circular processes, which correspond to changes in the equilibrium of chemical elements in various Earth's shells over a fairly long geological time. They correspond to at least 99.8% (by weight) of all matter in the Earth's crust. The weight fraction of atoms not covered by circular cycles plays an exceptionally important role in the Earth's crustal mechanism. These circular processes can be traced from the biosphere to the basic (basaltic) shell and back. The area of the circular processes - the Earth's crust - in a vertical section does not exceed 130-140 km. The PT factor has caused the thermodynamic shells to exist. In the current work, covering the phenomena in the first scientific exact approximation, studying them in the biosphere, it is convenient to simplify our ideas and unite deeper geospheres in the study of circular geochemical processes. Vadose - elements and minerals of the biosphere, processes occurring in the biosphere. Freatic - phenomena occurring in the stratisphere and in the upper metamorphic geosphere. Juvenile - lower metamorphic and below. At the same time, it is convenient to distinguish: 1. circular processes corresponding to several such thermodynamic shells; 2. the same processes occurring within one thermodynamic shell. The first corresponds to deeper and more powerful phenomena than the processes concentrated in one such thermodynamic shell. Energy of geochemical circular processes. The most important fact in these primary cycles is the slow but constant circulation of chemical elements from one geosphere to another. This movement occurs as a result of geological processes, and there is a manifestation of energy, the source of which is different. The geochemical history of the chemical elements is largely based on the study of the laws of such atomic migrations. The first fact requiring definition is the area of the Earth's crust occupied by these migrations, reducible to reversible processes, in which, in their pure form, the chemical elements always return to their original state. Thus, there must not be any loss of chemical elements, they must always remain within the boundaries of the field in which these reversible phenomena occur.

There is a definite area of atom migration.

Л. Teyseran de Boer and R. Assmann - independently distinguished between - the troposphere and the stratosphere. They proved that there is almost no exchange of gas masses above (from the geoid level). The boundary is a layer of minimum temperature -100-980 C. Above and below this layer the temperature rises. This layer, as any gas mass in complex equilibrium phenomena, penetrated by radiation of different changing energy, fluctuates within certain limits from geoid level and lies higher above the equator of the Earth than above the pole. The gases of the stratosphere above are very independent of the motion of matter on the earth's surface, and although there is an exchange between the matter of these high regions, the matter of the stratosphere and the earth's surface, this exchange is extremely slow. Undoubtedly, over geological time, it will not be an imperceptible quantity. In the troposphere are quantitatively felt echoes of geochemical reversible processes. The thickness of the troposphere differs at the equator and pole, over continents and oceans, is somewhat variable and should range from 9 to 13 km. It is more powerful towards the poles. One can consider that lower boundary of the Earth crust is isostatic surface of the planet, which separates its surface area, clearly heterogeneous from gravimetrically, i.e. heterogeneous by specific weight of its constituent rocks, from the gravimetrically homogenous deep zone. If we exclude the hydrosphere with specific gravity of 1.03, differences in specific gravity of large parts of the surface part of the planet would not exceed 5% in total, rising sometimes to 10%. Everything indicates that the deepest parts of the Earth crust are much more homogeneous: their specific gravity is equal on the same level (counting from the Earth`s centre or from the geoid`s surface), i.e. we have a regular change of the shells with equal specific gravity while moving to the Earth`s centre. Gravimetric survey showed that isometric surface in bowels of the planet is located in depth from 59 up to 96 km. No geological phenomenon brings matter not only from areas below this surface, but also from much shallower depths. Voids in rocks may exist down to a depth of 12-15 km. It becomes clear that the break in the properties defining the granite geosphere of the Earth's crust lies much closer to the surface, not deeper than 16-20 km. This lower boundary is very irregular, given by the topography of the planet; it approaches the bottom of the ocean, sinking heavily beneath the continents. Most of the earthquakes begin at a depth of 8 km (R. Oldgem, 1926): these are episeisms: rarely do they reach the depth of 34 km (T. Ogawa, 1924): there are bathysseisms, they begin at the depth of 200 km, rise to 100 km and descend to 300 km (H. Turner). From the point of view of interest, only episems are important, but their importance from the point of view of exchange of substance of surface and deep parts of the Earth's crust is insignificant. This exchange ends in the granitic shell and captures the basaltic, underlying, mainly oceanic and adjacent land areas. The oceans have not been shifting for at least 1 billion years - since the Cambrian and earlier - then they may be considered as an outlet for the biospheric migration of atoms of the deep basaltic geosphere enriched with magnesium and iron. This feature of the biosphere's organization remains unchanged within geological time. Migration of atoms of circular processes reaches only the granitic geosphere. Everything that lies below this boundary is not reflected in the processes under study, in their modern scientific understanding. Nor is it reflected in the migration of elements if we go beyond our time and take into account the slow processes occurring in geological periods: the formation of mountains, shifts, movement of earth's masses. They all proceed within the same limits. Geochemical cycles of chemical elements are a superficial phenomenon. Inner part of the planet for them as if does not exist" [1].

The Earth's autoscillatory system as a stationary energy centre of the first kind

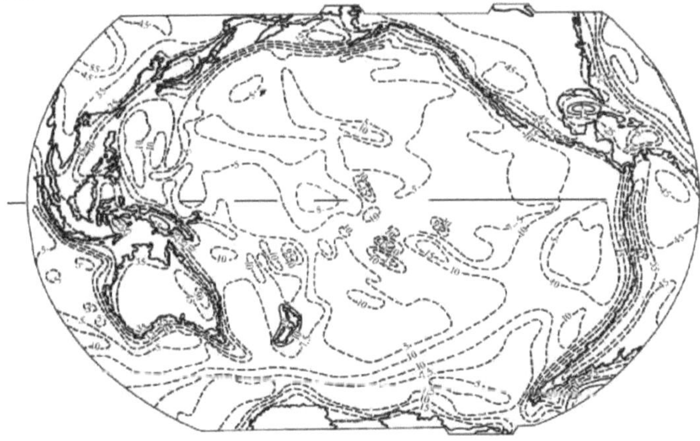

Model. Thickness (in kilometres) of the Pacific Ocean floor crust (by R.M. De Menitskaya)

According to V.M. Rarwalski, "a complex dynamic system is a set of objects that develops in space and time and is connected to each other in a certain way into a unified whole and consists of a large number of elements. A complex dynamic system has properties (emergence) which the objects and elements that form it do not have. A complex dynamic system is cybernetic when it has at least one control object (algorithm) that does not depend on the material implementation of the objects themselves" [5,7,10]. [5,7,10].

Studies have shown, M.M. Dovbich, N.F. Balukhovsky, that:

"cyclicity of geological processes, correlates well with cycles of certain astronomical phenomena related to rotation". L.L.Hudzinsky, studying seismicity of Elbrus region, concludes that "...processes taking place in active fluid-magmatic chambers are influenced by variations of gravitation field" [5]. [5].

The Sun revolves around the centre of the Milky Way galaxy. Its average speed is 828000 km/hour. One revolution takes about 230 million years. The Milky Way is a spiral galaxy. It is thought to consist of a central core, 4 main arms with several short segments. The solar system is located in the spiral subsystem of the galaxy, which has a high energy level - to the question of dividing space into low and high energy areas. The Earth system rotates on its axis, around the Sun and around the galaxy, while making quasi-sinusoidal oscillatory motions in the plane of the galaxy.

The Earth system is a deformed body of rotation that reflects the inhomogeneity of space. The structure of the planet is a deformed system (hierarchy) of blocks whose formation is associated with the existence of interacting stress fields of the Earth system and a wave of energy emanating from the core of the system.

The shape of the Earth system is close to the surface of an ellipsoid of rotation, whose equatorial radius is 6278.245 km and polar radius is 6356.863 km (the Krasovsky ellipsoid). The system can also be represented by a triaxial ellipsoid in which the difference between the large and small half-axes of the equator is 210 m. The core is bounded by a spherical surface with a radius of 3473.4 km. The difference between the equatorial and polar radii is 21.378 km, the average radius is 6371.2 km5 and the circumference is 40075.7 km, and the Earth's surface is 510000000 square km. The specific surface area is 29% terrestrial and 71% water. The partition between mantle and core corresponds to a depth of 2500-2900 km (which is respectively 0.608-0.545 of the radius, counted from the centre of the Earth as the basic point of reference). The boundary of the inner core is 4500-5000 km, i.e. 0.294-0.215. R.

"Auto-oscillation is the undamped oscillations in a system in the absence of a variable external force. The amplitude and period of the oscillation are determined by the properties of the system itself. In order for the oscillations to be non-durable, the energy entering the system must compensate for the loss of energy by the system. The values of the vibration amplitude at which the losses are compensated for in total per period are stationary; the vibration amplitude is determined by the properties of the system itself. When the oscillation amplitude is less than stationary, the inflow of energy exceeds the losses, therefore the amplitude increases, reaching a stationary value - self-excitation of the oscillations of the system occurs. At amplitudes greater than the stationary ones, the energy loss in the system exceeds the energy input, so the amplitude decreases, reaching a stationary value.

There are three basic elements in self-oscillating systems: the oscillating system; the energy source; the device that regulates the flow of energy from the source to the oscillating system" [5,10].

There is a temporary lag of the hydrothermal ore-forming process and localisation of minerals of all types, in fracture-breccia, all morphological types of structures - flat, steep-dipping, tubular, flexural, etc., in which the following occurs: gold, uranium, strontium, mercury, oil, gas, gas condensate and others (deposits of Central Asia, Western Siberia).

This circumstance is explained by the difference in mass-flow migration velocity - fluid, and the velocity of the energy wave, under the influence of which the cyclic-directional structuring of the geological space of the Earth system takes place.

The process of structuring is of particular importance in the formation of regional weakened gentle energy reservoirs, which are formed under the influence of energy waves of the Earth's auto-oscillation system. In such energy reservoirs hydrocarbon deposits are formed in the era of neotectogenesis from the Upper Cretaceous and independent of the shielding rocks (Middle Priob'ye - "unstructured" deposit). The hydrocarbon-oil-oil-condensate-gas zoning is clearly manifested in the WSP. It should be mentioned that gas saturation of oil took place as a result of overpressure from the mantle. Periodicity and discrete location of HC deposits in space and time is manifested.

The Kochbulak ores are gold, (hydrothermal process), localised, in pre-ore morphostructures of secular, declivity. The ore-calcifying structures are located periodically and discretely. An analysis of the localisation conditions of the minerals indicates that they are associated with zones of increased permeability, irrespective of the composition of the host rocks.

The value of 21.4 km, causes the limit value, the amplitude of the vertical displacements along the Earth's radius.

The real limit value of the hypsometric spread recorded at the surface of the Earth is 19 km.882 m. It is defined by two extremes: the extreme altitude of mountains equal to 8848m and the greatest depth of the ocean floor (Mariana Trench) equal to 11034m. Comparing the values of the range of possible surface elevation changes (21.4 km) and the real limit value of hypsometric range - the difference between them is 1.5 km (7%) - a constant value of losses due to friction in the Earth's auto-oscillation system. The damping decrement of the Earth's auto-oscillation system is very high - 0.93 (the Earth's system efficiency) [10].

The real limiting value of the hypsometric spread recorded at the Earth's surface is 19 km 882 meters. Questions arise: what is the minimum wavelength within which an amplitude equal to 19.9 km is realized and what are the dimensions of other waves generated by the Earth's auto-oscillation system [10].

The rotating Earth, a self-oscillating system, has a set of natural oscillations which produce a single all-Earth system of standing waves, each a generator and a tuning fork, capable and ready for resonance. When private oscillatory systems arise in the Earth's interior, interference inevitably occurs. If the periods of the local waves coincide with one of the waves, there is resonance. The emergence of Earth-wide standing wave zones is the main shaping mechanism for planetary structures. Harmonics arising on the basis of universal standing waves turn out to be the main mechanism forming regional geological structures. The resonance occurring as a result of the interference of the waves generated by the universal and regional sources leads to the formation of local structures. In other words, the system of terrestrial standing waves and regional waves and harmonics formed on their basis, as well as resonance of waves generated by them and regional waves create ordered interference lattices, on the basis of which tectonic dislocations - plicate and disjunctive structures appear.

1. The level of energy expended on oscillatory movements in each particular areal determines not only its size but also the size of the tectonic structures formed and the amplitude. Tectonic dislocations formed in separate geological regions have a systemic character and reflect both general terrestrial properties and regional features. The formation of structures of local significance is determined by the depth of the occurrence of oscillatory movements.

Taking as a first approximation the Earth's vibrational system as a string whose length is equal to its diameter, it is obvious that the deeper the excitation source is located, the poorer its overtones are and the stronger the main structure-forming tone is manifested. The Earth's self-oscillating system is non-linear, as the frictional force in it is constant for each level of its dynamic equilibrium and is directed in the opposite direction to its velocity. In such a situation, the system can only perform a certain number of half oscillations and its frequency spectrum goes out, in what is called the stasis band. "In the papers known to the author, (Bogatsky 1986) no models are published which allow us to estimate periodicity and wavelengths generated by a nonlinear self-oscillating system. Proceeding from the notions of ball symmetry, basics of wave mechanics and based on Fermat numbers:

$$N = (^{22})^{n+1}.$$

V.V. Bogatsky proposed a model for calculating the natural vibrations of the Earth. Operating on the notion of wave number "K" and Fermat numbers, which as proved in 1976 by C. Gauss, characterize the regular inscribed polygons if the Ferma number turns out to be prime. Prime Fermat numbers occur at n=0, 1, 2, 3, 4, and are respectively 3, 5, 17, 257, 65537. For the Earth's auto-oscillatory system the half-wave lengths of its principal overtones - harmonics should be multiples of 1/3, 1/5, 1/17, 1/257, 1/65537, with the principal half-wave length (tone) - /1/.

Thus, quantization of waves in the Earth's self-oscillatory nonlinear system occurs both by frequency within each subsystem and by the decrement of attenuation, by which the number of subsystems is specified. Based on the calculations, the nonlinear self-oscillatory Earth system should have six levels of hierarchy.

According to V.V. Bogatsky. 1986:

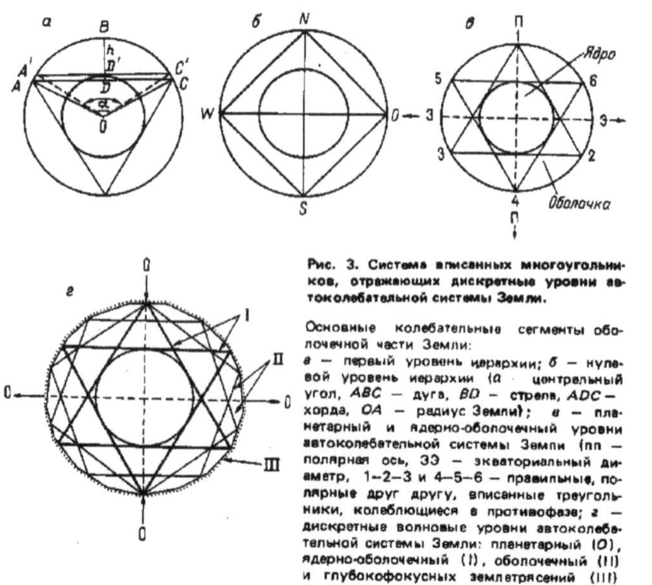

Рис. 3. Система вписанных многоугольников, отражающих дискретные уровни автоколебательной системы Земли.

Основные колебательные сегменты оболочечной части Земли:

а — первый уровень иерархии; *б* — нулевой уровень иерархии (*a* · центральный угол, *ABC* — дуга, *BD* — стрела, *ADC* — хорда, *OA* — радиус Земли); *в* — планетарный и ядерно-оболочечный уровни автоколебательной системы Земли (пп — полярная ось, ЭЭ — экваториальный диаметр, 1—2—3 и 4—5—6 — правильные, полярные друг другу, вписанные треугольники, колеблющиеся в противофазе; *г* — дискретные волновые уровни автоколебательной системы Земли: планетарный (0), ядерно-оболочечный (I), оболочечный (II) и глубокофокусных землетрясений (III)

Hierarchy level 0 (planet Earth) - basic tone 1; Hierarchy level I - overtone 1/3; Hierarchy level II - overtone 1/5; Hierarchy level III - overtone 1/17; Hierarchy level IV - overtone 257; Hierarchy level V - overtone 1/65537.

Geomorphological realisation of the amplitude above (+) and below (-) the geoid level: **hierarchy level** (y.i.): **/0/** - (-11060) (+8848); **/1/** - (-4000) (+3200); **/II/** (-2500) (+1800); **/III/** (-700) (+800); **/IV/** (-46) (+27)); /V/ (0.17) (+0.13). The main Earth-wide standing wave of the Earth as a planet is realised in the form of continuous rising or falling, which surface is inclined at an angle of not more than fifteen minutes, which corresponds to a relief change of 3-4 meters per kilometre.

The planet-wide standing wave zones are a system of independent emitters, each generating waves of lower amplitude but higher frequency - their own harmonics.

Planetary standing wave zones are regional wave generators. As a result of interference of planet-wide waves of different hierarchy levels, as well as interference of such waves by excited regional oscillators, resonant fields arise, resulting in the formation of contrasting local structures.

Local structures have a limited range of propagation, but the area of their propagation is determined primarily by the area of propagation of the resonating half-waves, i.e., essentially the vibrational subsystems that create them. Consequently, the size of local structures may vary widely, as it depends on the parameters of the waves that create them.

In the same region, local structures of different sizes can arise, from large to small. The notion of local structure is inherently relative; it is determined not by the size of the structure, but by its position relative to the generating (specifying) vibrational subsystems. It is important to note that the local structures are relative with respect to their generating vibrational subsystems as well as to each other, because each of them appears only on the background of another relatively larger one.

2. All this *defines one of the conditions for contrasting local structures - polarity by sign (oscillation phase)*. The contrast can also be expressed by the stress potential, which is externally ascertained by the change in occurrence conditions. The latter represents either a qualitative change in occurrence (e.g., folded forms - disjunctive forms) or a quantitative one (e.g., abrupt change in dip and/or strike angles). Consequently, *the contrast of local structures within a certain areal emphasizes both their specificity and isolation (10)*.

In September 1905, A. Einstein's article "Towards the Electrodynamics of Moving Mediums" appeared, in which the provisions of the special theory of relativity were first formulated. The relationship between mass and energy:

$E=mc2$, (1)

where, E is the energy of the system, m is its mass, c is the speed of light.

Energy: (E), units, SI system-(J), GHS system-(erg).

established by Einstein on the basis of the theory of relativity, plays a decisive role in questions related to the use of intronuclear energy. The ideas of relativism, together with the ideas of quantum mechanics, have advanced the understanding of the nature of elementary particles. Based on the requirements of relativistic invariance, the basic quantum equations of motion of elementary particles were derived; the relativity theory relations determine the transformation of some elementary particles into others.

Newton's law of universal gravitation:

$F=G (m M/ R2)$ (2)

F is the force of interaction of bodies, M and m are the masses of interacting bodies. Here G is the gravitational constant, equal to about 6.6725*10-11 m3/(kg-c2).

When m1=m2=m we have

$G=Fr2/m2.$ (3)

This formula shows that the gravitational constant is numerically equal to the mutual gravitational force of two material points having a mass equal to one unit of mass and located at a distance equal to one unit of length from each other. The numerical value of the gravitational constant is established experimentally. This was first done by the English scientist Cavendish using a torsioning dynamometer (torsion scales).

In the SI system, the gravitational constant is

$G = 6.6710{\sim}11$ Nm2/kg2. (4).

Consequently, two material points with a mass of 1 kg each, 1 m apart, are mutually attracted by a gravitational force equal to 6.67-10-10 N.

When we study the attraction of bodies according to the law of universal gravitation, we encounter the gravitational interaction between bodies. This interaction is one of the fundamental interactions that exist in nature. It takes place at a distance, without direct contact between the interacting bodies. Gravitational interaction between bodies, as described by the law of universal gravitation, takes place through a gravitational field (gravitational field). At each point of the gravitational field, a body placed there has a gravitational force proportional to the mass of that body. The gravitational force is independent of the medium in which the bodies are placed.

The force of gravity is directed towards the centre of the system, causing the planet to take on a spherical shape, with the primary abiogenic fusible, volatile elements and their compounds being displaced into the Earth's crust of magmatic origin.

The works of M.V. Petrovsky, A. Kaye, P. Tricar, have shown that "tectonic structural forms formed in the Earth's crust are displayed in the form of certain landforms. Epeirogenic processes are expressed in periodic deformations, which are caused by the passage of waves generated in the Earth's interior. The oscillations of different orders occurring in the Earth have been established by precise instrumental measurements. The summation of the oscillations leads to the phenomenon of resonance" [5].

According to V.V. Bogatsky [1986], "Zones of increased deformation separate relatively quiescent areas. They are also reservoirs of magma, fluids and hydrothermal solutions. The size of zones of increased deformation is very different, and within each zone of increased deformation lower order zones separated by relatively quiescent areas can be distinguished. Given this multistage of deformed zones, a single pattern can be made for all tectonic relationships, from planetary to local. The geological law, which is formulated here, is a reflection of two physical laws:

1. Any deformation of a solid and ductile body leads to its division into zones with a predominant deformation concentration and into weakly deformed blocks separating these zones, and such zones and blocks may contain separate

zones and blocks of a lower order. The lowest order of high deformation zones are some of the crystal lattices. The higher order depends on the dimensions of the deformed body. During deformation, new zones arise and the old ones harden, but with increasing strain they can come back to life.

2. High deformation zones are characterised by an increased degree of permeability to magma, fluids, gases, hydrotherms and stress waves" [10].

The formation of the system: vaulted uplift - Benioff zone - oceanic depression is associated with the division of geological space by a zone of intense permeability, which has a high energy potential.

The separated regions have not only different energy potentials, but also different degrees of tectonospheric permeability, which influenced the formation of the granite-metamorphic layer of the Earth system. The energy wave emanating from the core region also contributes to the expansion process of the Earth System. Deep fault systems control the migration of matter within the Earth system, the location of energy sources, and the formation of the architecture of the tectonosphere.

"The average density of the Earth is 5.52 g/cm3. Sedimentary rocks are 2.4-2.5 g/cm3; granites and most metamorphic rocks are 2.7 g/cm3; basic igneous rocks are 2.9 g/cm3. The average density of the Earth's crust is 2.8 g/cm3.

From a comparison of the Earth's rate of rotation and flattening with seismic wave velocity data at different depths and partitions within the globe, the following density values are now considered most likely:
- in the upper mantle roof - 3.1-3.5 g/cm3;
- at a depth of 1,000 km - 4.5 g/cm3;
- at a depth of - 2900 km - 5.6 g/cm3;
- in the roof of the core - 10.0 g/cm3:
- at the centre of the Earth - 12.5 g/cm3." [V.V. Belousov].

"The average value of heat flux is 1.5*10-6 cal/(s cm2 · the most common value is 1.1*10-6. cal/(s cm2

Significant local variations in these values are observed. The fluctuations correlate with modern endogenous zones as well as the degree of expression of the asthenosphere: in those zones where the asthenosphere is stronger, the heat flux is more intense; where the asthenosphere is weak, the heat flux has the lowest values.

In zones of weak orogenesis in place of Paleozoic geosynclines (Ural) the flux intensity rises to 1.5 in the same units. But in the Tien Shan, where strong recent tectonic activation is observed and where the asthenosphere is well expressed, the heat flux increases to 1.8. Even higher values of heat flux in zones of riftogenesis - 2.0 and zones of modern volcanism to 3.6." (V.V. Belousov,

1975) [9].

"The simultaneous manifestation (according to V.V. Belousov, 1975), on the surface of continents of different endogenous regimes, "indicates the heterogeneity of the Earth thermal field: at the same time heat flows in different places differ in their intensity, hence heat flows change their intensity both in space and in time" [9].

"This fact indicates the existence of a single control mechanism under the influence of which the system and objects in its geological space evolve. This circumstance, gives an opportunity of wide application of method of analogy in geology.

Regularities in the structure of the Earth's crust blocks are manifested at the regional level, which is very important for solving zoning and forecasting issues" [5,7].

S.V. Starchenko.

IZMI RAS, Troitsk, Moscow Region.

"A comparative energy analysis of the main known physical phenomena determining the internal currents and the magnetism of the Earth's core generated by them in a reference frame rigidly tied to the Earth's mantle was carried out. The energy impact of a particular phenomenon was evaluated by the efficiency of its impact and what maximum and minimum contribution it is able to make to the kinetic and magnetic energy of the core S.V. Starchenko, 2009". [5,7].

"Ph.M., Professor S.V. Starchenko testified that:

1. The core magnetism is most effectively and powerfully influenced by currents generated by gravitational-chemical processes, which are predominantly represented by gravitational mass differentiation due to the growth of the Earth's inner core as heavy components are deposited from the cooling outer core.

а) . Gravitational and chemical processes are converted into kinetic and magnetic energy almost without loss.

б) . During gravitational-chemical processes, several TW (1 TW=1012 W) are emitted. A power of the order of 0.5 TW is sufficient to generate the observed magnetic field of the Earth and to maintain the magnetic field hidden in the depths of the core.

в) . The thermal impact is considerably less effective. Its total power is 10 TW (in the core), but less than 20% of the thermal energy is transformed into hydromagnetic energy.

2. The thermal energy at the core-mantle boundary is 6 TW, of which 1 TW is converted into the hydromagnetic energy of the core.

3. The effect of structural factors, as well as external factors such as the Moon and the Sun, on the hydromagnetic dynamics of the core is very small and

difficult to assess."

These researches of S.V. Starchenko, allow to solve an inverse problem and to establish the reasons and consequences of structural-material transformation of Earth system and concentration of mineral raw materials under the influence of energy wave with capacity 10 -13 TW and to draw the following conclusions [V.N. Ustiantsev] that:

This region: the core-mantle boundary, is the zone where the energy wave occurs, under the influence of which the material-structural transformation of the Earth system occurs. Juvenile postmagmotic solutions, are the retarder of "nuclear" reactions occurring in the zone of the system: core - bottom of the mantle (melt of the D11 shell). The genesis of the whole range of elements that take part in the formation of mineral deposits, which are localized in blocks of the Earth's crust, is connected with the post-magmotic (juvenile) solutions.

Due to the fact that:

Of all known natural phenomena, the system properties of the energy wave are able to structure the Earth system space with the manifestation of patterns of deposit placement in blocks of the Earth's crust. The deposits are located in the blocks, obeying a certain law, that is, the complementarity of the energy wave system properties is shown. As shown in work discreteness, periodicity of placing of deposits of mineral raw materials is shown.

Total power of energy wave coming from the region of core and bottom of lower mantle is about 10 to 13 TW. That is, under influence of energy wave with power from 10 to 13 TW, there is a structural-material transformation of the Earth's self-oscillatory system.

This position is fundamental to understanding the architecture of the Earth system and the mechanism of the processes taking place in its space.

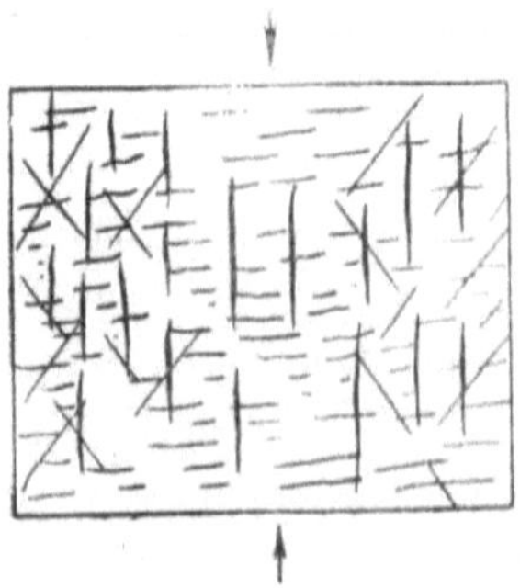

Cracks occurring in a layer of paraffin under compression. (by W. Mead).

Ellipsoid deformation

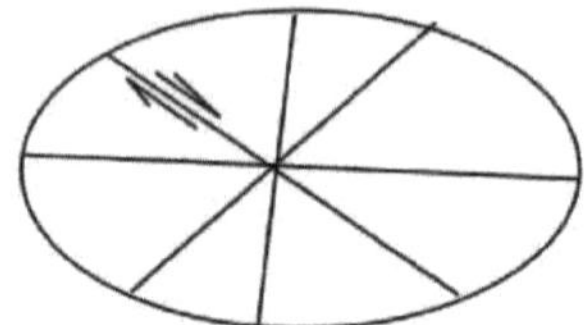

The ellipsoid of deformation. The four main strike-slip directions of tectonic faults. V.N. Ustyantsev, 1989).

The concept of the deformation ellipsoid was introduced into geology by Becker in 1893. Becker used his theory to explain patterns in the location of fractures and the genesis of cleavage. Becker's hypothesis was not immediately accepted, but only after observations in different parts of the world had established a clear pattern in the location of fractures. Since that time, geologists have become interested in the applied theory of elasticity (resistance of materials), and later also in the theories of plasticity and strength" (V.M. Kreiter) [5,11].

The regular arrangement of the structural elements in the Earth system space

The dependence of the internal structure of geosynclinal (folded) systems on the spatial position of deep faults at intervals of 10, 20, 30, 40 km. from each other. I.e. discreteness, on the one hand, and interrelation of these structures with each other, as well as straight-line faults, intrusions, fracture zones, lithoformational changes and morphological changes, on the other hand, are shown by the example of the western part of the Altai-Sayan folded area by M.A. Churilin. He also proved the discreteness of area (isometric in plan) structures associated with a decrease in the radius of arc-shaped geological boundaries, expressed by zones of intense tectonic deformation, including deep faults within folded systems, from ancient to young. These discrete elements are related to each other through coefficients ^{2} *and*.

The ***periodicity of localization*** of ore areas was pointed out by G.L. Pospelov. Analyzing the regularities of the location of magmatic iron ore deposits in the Altai.

of the Sayan folded region, he showed that:

"The intersecting structures, consisting of linear systems of structural elements, collectively form a geotectonic grid, which is determinant for the location of iron ore belts, and often, for the location of ore clusters and individual ore fields. Such grids have a defined step in the latitudinal and meridional direction (160, 80, 40, 20 and 10 km)" [10].

Among the works devoted to the quantitative analysis of the structural elements of the Earth's crust, the publication of V.N. Semenov and V.V. Brongulev. In their opinion, "the size of folds may serve as the most general and simple criterion for their subdivision. In order to construct a better scheme of scale classification of folds we should take as the main parameter not their area but the length of large axes. V.V. Bronguleev has established that the dimensions of fold groups represent an ordered series obeying the sequence of powers of two. He conventionally accepted folds with long major axis from 0.5 to 1 km as lower order forms" [10].

G.I. Leontiev concluded "about a single structural-hydrographic series of morphometric indicators of geological structures and geomorphological elements" (river valleys) [10].

The regularities of the structural series are explained by tectonics.

G.L. Pospelov noted "regularities in the geometry and dimensions of fractures, which led to the concept of planetary fracturing" [5,10].

According to Y.A. Meshcharikov "the existence of geowaves is connected with geometrical regularity of morphostructures location and reflects some general planetary regularities, including the general geometrical regularity of the Earth's figure. The meridional-latitudinal arrangement of geowaves expressed in relief, is connected with the position of the Earth rotation axis".

S.M. Kravchenko shows that in the "regions of the Aldan Shield, Oslo graben, Caucasus, Kamchatka, and East Africa, the distance between volcanic centres varies regularly. The main maximum distance is slightly greater than 8 km and coincides or is close for various regions; it corresponds to the average diameter of 114 ring complexes (according to Billings, this value is 8.3 km), the other two maximums are multiples of each other and are equal to 4.8 and 12.5 km, i.e., the series 4.8; 8.2; 12.5 km is outlined. *The establishment of* block *parameters* determining the localization of volcanic centres allows to predict conditions of localization of extrusive, intrusive bodies and deposits connected with volcano-plutonic complexes.

A generalized idea of the distribution of volcanoes along latitudinal zones also allows us to establish a periodicity with a step of [200] (V.V. Bogatsky), and a similar pattern is outlined in the meridional direction.

In 1968, B.I. Suganov discovered "a discrete periodicity in the location of magnetite deposits in southern Middle Siberia" [10].

M.A. Churilin outlines "links of discrete structures with metallogenic and ore fields, nodes, areas, including for intrusions of the central type".

The wave process is well documented in the coal fields. For the central area of the Donbass, V.N. Volkov, established waves with half-wave lengths of 7.6-10; 1.9-2.7; 0.35-0.45 km.

K.V. Gavrilin has observed a dependence for the coal seams of the Kansko-Achinsky basin, where the half-wave is 6-8; 2-4; 0.5-1 km.

"The Earth wave is essentially an enveloping curve that hugs the periodic alternation of maxima and minima in steps of 10o in the zone between 40o N and 40o S. And step of [20o] is characteristic for higher latitudes. The periodicity of maximums is revealed, the connections of which are located in 20o, 40o, 60o .

The similar periodicity in the density of volcanoes (orthogonal network), indicates the same periodicity of manifestation of stress fields in the Earth's crust - 20o step changes of the intensity of stress fields, covering the entire spherical surface of the Earth" [V.V. Bogatsky, 1986] [5,7,10].

Studies by P.S. Voronov have shown that "the development of tectonic processes during the Alpine and Hercynian tectogenesis proceeded according to the same laws, as they depended on the same causes".

P.S. Voronov emphasizes that latitudes of 42 and 44 degrees are favourable, especially 43-42 degrees for HC localization.

A number of depth structure features are revealed *by A.A. Borisov 1972*, by recalculating the anomalous field of magnetic anomalies at different altitudes. The recalculations for 100-200 km altitudes established a sub-latitudinal anomaly: the fields of positive anomalies are traced along parallels 70, 56 and 42 degrees, and negative anomalies along parallels 65 and 50 degrees [8].

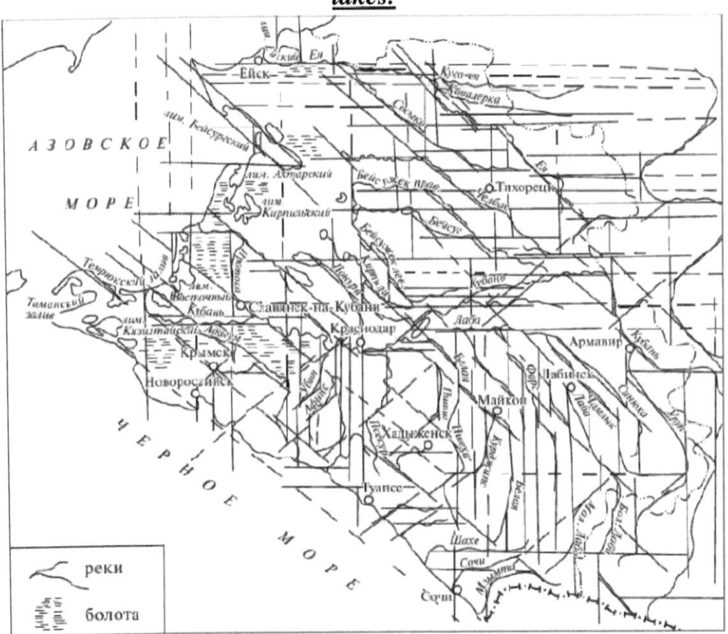

Western Caucasus. The river network marks the faults. Four main on the board (V N. Ustyantsev, 1986)

Analysis of the maps, at different scales, shows

- the geometric regularity of tectonic faults in the crust and upper mantle;
- tectonic faults control Earth system objects (volcano-plutonic complexes, crustal geomorphological elements, rivers, lakes, seas, oceans, continents and mineral deposits of various types and kinds);
- This fact points to the existence of a single controlling mechanism for the formation of tectonic faults in the Earth system;
- The zones of discontinuous systems (hydrocarbon-bearing) are distinguished, these are latitudinal and meridional strike-slip faults;
- are NNW strike-slip faults and NE strike-slip thrusts, which have developed, for the most part, since the Proterozoic;
- thrusts contribute to the formation of structural "traps" of minerals;
- zones of tectonic fracture systems girdle the Earth system;
- hierarchy of tectonic faults, forms a "rigid controlling system of geological processes" of the Earth's tectonosphere;
- patterns reflect the systemic properties of the Earth's envelope region and

also indicate a genetic link between tectonic faults and folded structures and the upper mantle;

- The manifestation of patterns of tectonic disturbance in the Earth's crust and mantle is an indication that the predictions are correct;

- Deep fault systems are oscillatory loops that control the migration of matter through the Earth system, the location of energy sources and the shaping of tectonospheric architecture.

As shown by simulations (Garat I.A. 2001), "energy of elastic wave generated by local generator increases permeability of weakened zones and faults by two orders of magnitude, while porosity increases by five times" [5]. This fact explains high degree of permeability of zones of deep fault systems and their high energy.

Weakened, easily erodible fault zones of all types, the geometry of coastlines - rivers, lakes and the sea - mark faults that are accompanied by resonance-interference, permeable zones - which are collectors of HC and other mineral associations (the fault is a second kind energy wave generator, developing syngenetically inherited).

It is important to note that the hydrographic network, the shoreline geometry of seas, lakes, - is recorded by topographers - ***instrumentally.***

In other words, what we get is a cost-effective, high-precision geological survey, so important in mineral prospecting and exploration.

The Earth's autoclave system is compressed along its short axis by 21.4 km, which leads during its rotation to periodic opening of tectonic faults of discontinuous type, mostly of latitudinal extent. Thus, centrifugal forces of rotating Earth system, promote formation of hydrocarbon deposits [5,7,10].

A meridional thick band of hydrocarbon deposits, from Kalmykia to the Arabian Plateau, has been identified. The hydrocarbon deposits are controlled by latitudinal and meridional tectonic faults. Meridional oil and gas bearing zones are identified in the west of the Caspian and in the east of the Azov Sea.

The meridional and latitudinal system, a fracture type, is not as pronounced as the diagonal thrust-slip dynamo; it was maximally developed after the formation of the sedimentary-metamorphic layer (early Riphean-Vendian). "M" and "S" zones are hydrocarbon-bearing.

Meridional and latitudinal, diagonal zones of deep fault systems, well expressed by the geometry of the shorelines of rivers, seas and oceans.

The fact that deep faults are pronounced on the surface, despite 20-30 km of sediment, is a vivid manifestation of the fact of the Earth's self-oscillating system (faults are the generators of energy waves of the second kind).

A.I. Suvorov has established that "north-eastern faults are characterized by thrusts and north-western faults by shear faults, which articulate at right or obtuse angles and form pairs of faults - "dynamopairs" [13,17,18]. This is explained by the rotation of the Earth system around its short axis in a certain direction. NE strike-slip thrusts contribute to the formation of structural "traps"; shear "spreads" (buddies) the deposits, in some cases, the shear may play a positive role.

The regular arrangement of the structural elements in the space of the Earth system.

The method of geometric geoprocessing, - is very reliable and accurate, as the cosmogenic factor, which is responsible for the patterns of location of space objects, and hence the structural elements of these objects (location of ESS, faults, deposits), in the Earth system

The main factors in the formation of tectonic faults (V.N. Ustyantsev):

1) The division of objects in geological space by the zone of intense deformation into regions **of** high and low deformation occurs irrespective of the shape of the object and the mode of its movement, but as a result of the action of gravitational forces;

2) during rotation - due to the centrifugal forces of the rotating system;

3) . the presence of a global, regional and local, stress field, the unloading of which led to the formation of faults;

4) a wave energy transmission mechanism that operates continuously in time and space.

5) . If the effect of external forces exceeds the rock strength limit, rock integrity is compromised - fractures or shear cracks occur, and crack development occurs in areas of maximum lithostatic stress.

Due to the fact that faults are primary structures, they are located linearly and have an end-to-end development, in relation to other tectonic structures, which allows **the** successful application of different geometric methods for forecasting purposes.

Deep fault systems control the migration of matter through the Earth system, the location of energy sources and the shaping of the architecture of the tectonosphere.

The principles of P. Curie:

"When certain causes cause certain effects, the elements of symmetry of causes must show up in the effects they cause."

"When a certain asymmetry is found in any phenomenon, the same asymmetry must also be found in the causes which give rise to it.

"Positions the opposite of these are wrong, at least practically; in other words, the effects may have a higher symmetry than the causes that caused them."

The primary importance of these provisions, very perfect in all their simplicity, is that the elements of symmetry in question apply to all physical phenomena without exception.

The symmetry is manifested in the geometric regularity of the zones of tectonic fault systems in the earth's crust.

Theorem of I. R. Prigozhin (1947), thermodynamics of non-equilibrium processes:

"under external conditions preventing the system from reaching an equilibrium state, the stationary state of the system corresponds to the minimum production of entropy".

"Synergetics explains the process of self-organisation in complex systems as follows:

The system must be open. A closed system, in accordance with the laws of thermodynamics, must eventually reach a state with maximum entropy and stop any evolution" (I. Prigogine).

"Self-organisation is inextricably linked to wave processes" (I. Prigogine).

"In any open, dissipative and non-linear system, self-oscillatory processes, supported by external sources of energy, inevitably occur, resulting in self-organisation" (Prigogine).

The process of mineral deposit formation is anti-entropic. The system of formation of mineral resources is open, due to the presence of tectonic faults in the Earth's crust. Thus, the main factor of deposit formation is tectonic faults. That is, tectonic faults control mineral deposits.

Of all known natural phenomena, the system properties of the energy wave are able to structure the Earth system space with the manifestation of patterns of deposit placement in blocks of the Earth's crust. Deposits are located in the blocks, obeying a certain law, i.e. complementarity with system properties of the energy wave is shown. As shown in work discreteness, periodicity of placing of deposits of mineral raw materials is shown.

The above does not allow us to conclude that the Earth's auto-oscillatory system is disseminated.

"Stability of the processes of regional structure-formation as a planetary quality of the Earth system together with periodicity and discreteness of the same regional structures testify to the fact that the main properties of the geological structures of all hierarchical levels reflect the unity of the universal planetary mechanism creating them. Such mechanism is the Earth autoscillatory system, generating waves of different length stresses, which are determined by peculiarities of its structure" [5.7,10].

All structural elements, in the geological space of the Earth system, are arranged in a regular way so that the dynamic equilibrium of the system is not disturbed (manifestation of cybernetic properties), which indicates the existence of a mechanism regulating geological processes, which is an auto-oscillatory system.

The spatial distribution of tectonic faults in the earth's bark

The **spatial distribution** of tectonic faults in the Earth's crust is one of the most important factors determining its architecture and, consequently, the external geomorphological shape of the planet, as elements of geomorphology reflect the internal structure of the lithosphere and the processes taking place there.

Схема. Метод телесейсмической томографии (ПТБ) (Л.П. Винник, В.Н. Устьянцев)

On the tectonic deformation plan of the block and the inherited evolutionary development of its structure in space and time.

The teleseismic tomography method.

The teleseismic tomography study assumed that the lateral heterogeneity is concentrated in the layer from the Earth's surface to a depth of 300 km. The strongest velocity inhomogeneities were found to be directly beneath the Earth's crust. The strongest decrease of longitudinal wave velocity in the central Tien Shan is about 3% of the average value but the algorithm used provides data smoothing and the real amplitude of velocity variations can be twice as large. In the upper mantle of hot spots an abnormally low wave velocity is observed [by L.P. Winnick], indicating elevated temperatures at depths of up to 250300 km. Seismoanomalies at depths greater than 400 km have been detected, but their

27

interpretation in terms of mantle plumes is still hypothetical. Seismic methods confidently map descending mantle flows of cold matter, beginning in subduction zones near the Earth surface and sometimes reaching the D ^{11} layer, at the bottom of the lower mantle. The downward flows up to a depth of 650 km are marked by earthquakes, while the upward flows are aseismic. With the huge size of the plume head in the subcrustal layer, the diameter of the feeding channel in the lower mantle may be 100-150 km, - at the resolution limit of seismic methods.

This makes it necessary to apply a general geological analysis as much as possible, using extensive and varied factual material.

The study of the Tien Shan *by the receiving function method* showed that the difference between the hot spot of the central Tien Shan and neighbouring areas is also manifested in the crust structure and the nature of the transition from mantle to crust: the velocity of transverse waves in the crust of the central Tien Shan at 10-35 km depth is several percent lower than outside it, and the transition from upper mantle to crust occurs over a *wider depth interval.* The "fuzzy" coromantine transition may be the result of vertical intrusions of mantle material into the crust, and the reduced shear wave velocity may be the effect of elevated temperature or the presence of fluids of magmatic origin. By comparing geophysical characteristics (relationship between phase velocities and Bouguer anomaly, heat flux values, bulk wave velocities and anisotropy) with similar data for other regions (Canadian Shield and Japan), they found that the crust of Central Asia, has some intermediate density or material properties. Here, there is a slight increase of heat flow, anisotropy (in the form of inconsistency of cuts by Lyav and Rayleigh waves at lambda = 110 km from Bouguer anomaly is located below such dependence for shields and platforms, but higher than in Japan. According to these researchers, the anisotropy is caused by attenuated layers (waveguides) in the crust and mantle [L.P. Vinnik, 1998] [5].

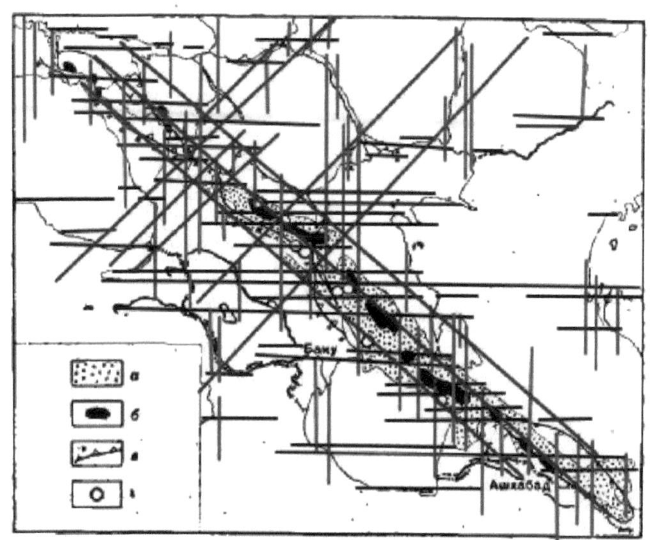

Магнитное поле Северо-Кавказского офиолитового пояса и эпицентры разрушительных землетрясений

а, б – магнитные аномалии интенсивностью 1 и 3–5 мЭ соответственно; *в* – контуры краевых прогибов; *г* – эпицентры разрушительных землетрясений

Дополнил: В.Н. Устьянцев, 2020

In many regions of Russia century changes of gravity are established. However, measurements from 1950 to 1965 on points Ashkhabad - Dushanbe, Tashkent, Alma-Ata, Balkhash established that there is no change in gravity, since the constancy of gravity is characterized by error ±0.008 mlg, then the possible magnitude of change below this figure. Employees of the Institute of Physics of the Earth, USSR Academy of Sciences, revealed an anomaly by calculating isostatic anomalies of gravity averaged over $^{10 \times 10}$ areas and caused by extensive density inhomogeneities at great depths. Against this background, regional anomalies with rather significant horizontal gradients up to 0.15 mLg/km are shown, their amplitude reaches several tens of milligal. The largest negative anomalies cover Central Asia with B=-1 density, layer thickness (anomalies) more than 500 km in Pamir-Alay; 350-500 km in Northern and Southern Tien Shan; Bukhara-Gazlin and Mari; 150-300 km - Fergana valley and Turan plate (IFZ RAS RF) [8,5].

The magnetic field of the North Caucasus ophiolite belt and the epicentres of destructive earthquakes.

The magnetic field of the North Caucasus Lineament and the epicentres and destructive earthquakes.

A, b, are magnetic anomalies of intensity 1 and 3, 5 Me, respectively;

d - epicentres of devastating earthquakes. (V. Eliseev).

Lineament of the North Caucasus, by V. Eliseev, 2018 [5].

Regional seismicity of space

"Regional seismicity of space gravitates towards the negative morphostructural elements of the Greater Caucasus (piedmont troughs and the developing depression of the Middle Caspian). The troughs are separated from the Greater Caucasus uplift by a suture zone, which is a hundreds of kilometres long, close (**30 to 100 km**) arrangement of deep faults penetrating the upper mantle and associated intracrustal faults.

The seam zone between the uplifting Caucasus and the dipping areas of the associated troughs runs along the southern flank of the trough zone and includes intense magnetic anomalies. The mutual correlation of magnetic and gravity fields has enabled the identification of a regional ophiolitic belt from the Crimea to the Predkopetdag Trough of Turkmenistan. The shallow occurrence of the roof of the magnetically disturbing masses (12-17 km) suggests the continuation of the fluid impact of the zone of intense paleomagmatic activity on the geodynamic regime of the overlying strata" (V.Alekseev) [5].

Recalculations of magnetic anomalies, for altitudes less than 50 km, show that two systems of Northwest Extension anomalies are clearly distinguishable. The Estonian-Pri-Caspian-Tajik system of anomalies is represented by predominantly negative anomalies from 0 to 1 Ma, with the axis of the system of lows extending through the Voronezh massif, the Caspian lowland, the lower Syr Darya, the Fergana valley and the Pamirs. Large positive anomalies (+1 to +2 Ma) are noted in the Beltaus crush zone, Bukantau and small ones in the Hungry Steppe and Farabsky uplift (+1 Ma). The northern boundary of the system is the Karatau-Fergana fault and the southern boundary is the Dnieper-Donets-Mangyshlak- Priamudarya fault.

To the south is the Baltic-Black Sea-Trans-Caspian system, represented by clearly visible and cumulatively arranged intense positive anomalies (up to 2-4 Ma). The axis of the system of maximums (MT) passes through the Krasnovodsk peninsula and Kopetdag. The endogenous energy flows lead to the formation of excess energy zones.

A number of depth structure features are revealed by A.A. Borisov, by recalculating the anomalous field of magnetic anomalies at different altitudes. The recalculation for 100-200 km altitudes established a sub-latitudinal anomaly: the fields of positive anomalies can be traced along the parallels 70, 56 and 42 degrees, and negative anomalies along the parallels 65 and 50 degrees. The Uzboy-Tarim zone - latitude [420] - can be traced to the Black Sea and controls structures in the Caucasus [8].

"In the zone of junction of epipaleozoic, more ancient plates, the main potential of oil and gas content is associated with the base of the sedimentary cover, in the area of crustal weakened horizon. The main potential of oil and gas content is associated with the processes taking place in the lithosphere and the upper mantle" (V.I. Popov) [4].

The middle massifs of the plate and platform region - mark the zones of oil and gas generation... The middle massifs of the movable belts area mark degassing zones due to the fact that they are not overlapped by the sedimentary cover in which mineral resources are localized. The role of buried structures of the Baikal tectogenesis cycle - NNW - Riphean is underestimated. During that epoch, the sedimentary formation - Blyne series - was formed, in which mineral raw materials were accumulated - Kopet-Dag, N. Caucasus, W. Siberia, Timan.

"Riphean geosynclinal formations are noted in the band from the Carpathian (P.S. Semenenko), through S. Caucasus (Khasaut Formation) to East Iran - Algonian murid series; as a result of the Late Baikal phase of tectogenesis - Gali cycle, crumpled into folds.

Another geosynclinal zone extends through the Timan and the eastern part of the Turgai Trough (Yu.D. Smirnov) into the Northern Tien Shan (V.G. Korolev). The main folding cycle is Baikal (Timan - Y.D. Smirnov, 1964). N.P. Kheraskov (1963) - one of the essential differences from the younger geosynclinal systems - broad marginal troughs of baikalids. The sedimentary formations of the Himalayan and Baikonur troughs, have the same formational composition (Blyne series, in which HC localization occurs). The Blyne series is also characteristic of the Bolshe-Karatau, Chatkal and Timan-Naryn troughs" [8].

<u>Illuminated structures:</u>

"Lineaments are the largest tectonic, global structures (Archean-Middle Proterozoic);

- linear, very long - **thousands of kilometres**;

- capacity **- up to 10 km;**

- are **spaced 50 to 100 km apart,** through-faulty, fracture-type disturbances. These faults are most prominent in the crystalline basement in the Archean-Proterozoic (before the change of deformation plan).

The lineaments are manifested in the roof of the granite-metamorphic layer. Extension is meridional-latitudinal - flexure-fracture tectonopair and diagonal - flexure-slip-slip-slip tectonopair (NE - flexure-slip; NW - flexure-slip).

Lineaments are well defined at the top of the granite-metamorphic layer and are often fluid-extracting structures. During eras of crustal degradation, faults open up and the reservoirs are fed with gas, an area of the upper tectonosphere that lies above the energy barrier - 0-12 km depths.

The lineaments of the four main directions, are traceable to a set of factors:

- along rectilinear landforms and geological contours, ancient and modern hydrological networks;

- by clear boundaries between landscapes, areas of denudation and accumulation, the occurrence of which is determined by endogenous causes. They are visible as light or dark bands (depending on the degree of reflection, emission, or absorption by rocks) on television infrared, radar multispectral photographic materials.

Lineaments sometimes coincide with the strike of deep faults.

In the areas where the lineaments intersect ore zones, the latter show elevated concentrations of the useful component.

Thus, isometric magnetic maximums are observed along the northeastern strike of the lineament zones (thrusts). In the lineament zone, gravity anomalies are systems of small, continuing each other gravitational steps, limiting in size anomalies of both signs. In some cases there is a reversal of isoanomalies, (according to N.A. Fuzailov, 1976), and with them local anomalies in the north-eastern direction " 8,13].

Isometric magnetic maximums indicate the flow of gas-saturated magma from the base of the Earth's crust along the lineament zone. The genesis of the gold, giant Muruntau deposit and giant methane deposit, Gazli, is associated with the junction of the lineaments of latitudinal and northeastern strike. Lineaments, are responsible for the process of formation of granite-metamorphic layer, with which genetically and paragenetically, the formation of large HC deposits is connected.

"Along the southern edge of the Kuramin massif, there is the South Fergana-Central Kyzylkum belt of basic and ultrabasic rocks (Carboniferous) - a "hot spot", *1200 km* long and *30 km wide" (I.Kh.* (I.Kh. Khamrabayev, 1975) [8].

Discontinuous ultrabasite bodies have also been identified in other areas, all gravitating towards deep fault zones. The basite bodies are pre-faulted. Covering of the basic rock structure and jespilites play the role of screens, i.e. they contribute to HC and oil generation and their migration in favorable conditions for their localization which are defined by PT factor (Amu Darya OB of hydrocarbons). In the Southern Tien Shan, according to the strike of deep faults, there are chains of hyper-basites (contacts - protrusive, which are considered as derivatives of the upper mantle [Khamrabayev, 1972] [8].

This zone of deep fault systems is associated with (correlated with) HC deposits, diamonds - oil, gas, gas condensates (Gazli deposit - methane with helium label - upper mantle). At the junction of the intersection of the latitudinal South Fergana lineament with the north-eastern lineament, there are giant deposits: the Muruntau gold ore deposit and the Gazli methane deposit.

"Mantle gas 'hung' in the upper layers of the Earth's crust, so no geologist has been lucky enough to find a volcano in the middle of a platform scattering diamond crystals around it. The kimberlite pipes that are found are only uncovered by erosion processes. For the prospector this means that there are many "blind" kimberlite pipes that do not come to the surface. Their presence can be recognized by detected local magnetic anomalies, the upper edge of which is located at a depth of hundreds, and if you are lucky - dozens of meters" (A.Portnov) [5].

It explains: *"Seismic data record the presence of zones of seismic transparency in the Earth's crust -* "zones of absence or significant weakening of reflecting and refracting boundaries", In such zones seismic waves travel with the least loss of energy. Their upper parts do not reach the surface and their upper ends can play a role of wave screens, where absorption and transformation (not necessarily thermal) of wave energy will take place" (G.B.Naumov) [5].

A.F. Grachev notices that, "the effect of subslattening is traced to much deeper mantle horizons than Moho boundary, as it has been established for ancient plume of trap province of Parana River. Here the low-velocity mantle anomaly, considered as a result of density deformation related to formation of a giant intrusion during solidification of mantle plume material, which is up to 300 km across, is traced to a depth of 500-600 km" [5]. [5]. These zones were formed under the influence of the Earth's auto-oscillation system. Deep fault systems, being oscillatory circuits, control migration of substance in the Earth system, location of energy sources and formation of tectonosphere architecture.

Among the most informative indicators of endogenous ore formation rightfully belongs mercury, which forms lithochemical, aqueous and atomochemical halos in soil and atmospheric air. Besides searches of ore deposits studying of halos of dispersion of mercury is effective at research of geothermal areas and zones of modern volcanic and tectonic activity, at an estimation of potential oil and gas bearing capacity of perspective structures. Thanks to the specific physical and chemical properties, mercury is the unique metal forming gaseous halos in ground atmosphere with the concentrations amenable to registration by instrumental optical methods for today. Systematic researches, have allowed to establish wide development of gaseous halos of mercury in ground atmosphere of mercury, gold-ore, rare-metal and other ore deposits. *The fact of existence of gaseous halos of mercury above a sea surface within regional tectonic disturbances (the Bering Sea) is established for the first time".* *(N.R.Mashyanov, 1985) [5,7].*

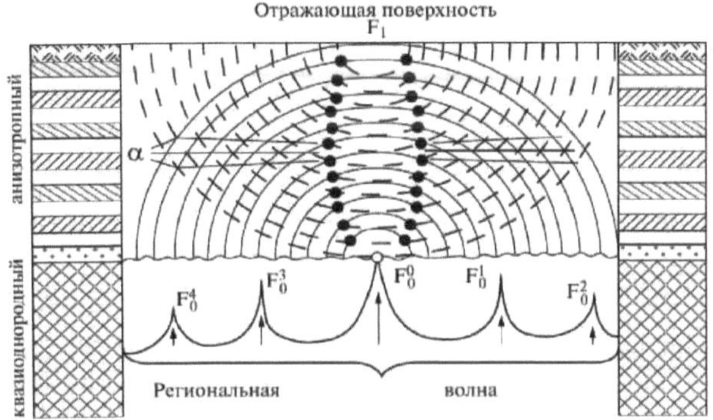

Model. Cross-section of a steel plate showing the system of spalling caused by a shock wave (from below) (by Reinhart and Pearson, 1973)

Model. "Hyperbolic and elliptical resonant surfaces as a result of interference of the wave front reflected from the Earth's surface generated by a deep source. Point contour in the centre of the figure: - interference ellipse, interference hyperbolic lines perpendicular to it (V.V. Bogatsky, V.N. Ustyantsev).

At the midpoint, between the excitation source and the reflecting surface, there is a resonant surface (V.V. Bogatsky) parallel to the earth's surface. If the distance between the wave source and the reflecting surface is large, the hyperbolic interference surfaces only take place up to a certain limit. When the interference surface corresponding to the stretching beams is able to produce discontinuous deformations, i.e. when the resonance energy is higher than the tensile strength of the medium, the fracture-break or weakened zones appear. If in the stratified section there are a number of reflecting surfaces, connected with the difference of physical properties of adjacent strata, then, besides the daytime, the uppermost reflecting surface, a system of intermediate screens - reflectors will appear, which will lead to a series of compression or tension surfaces and, consequently, to the formation of a series of flat fracture-breccia zones of fracturing and compression. If the wave-beam goes through some interval, on all its crests, corresponding to the beams, there will be impulses generating conditions in which the stretching and compression zones, subparallel to the Earth surface and to the interface surfaces of the strata composing the section, systems of plan-parallel "agreement" with the deposited sedimentary-volcanogenic strata, or oblique fracture-breccia zones and delamination zones [5,7] will be generated. $E=mc2$ (1) - A. Einstein's formula, indicates the equivalence **of** mass and **energy.** That is, the object of investigation: matter and energy. If this is true, then, energy can be transformed into matter, and matter, into energy.

"The persistence of the strike-slip of certain disturbance systems has been noted in many areas of the world. The F. Wening-Meynes grid of faults has emerged. Wening-Meynes,

M.A. Maideger, the geometric lattice of G.L. Pospelov and the ideal grid of planetary fracturing of the Earth's crust created by rotational motions [I.I. Chebotarenko, 1963].

At *the beginning of the 20th century, W. Hobbes* pointed out numerous examples of "geometric structuring" of the Earth's surface relief, in which rectilinear directions predominate. In the 30s of the 20th century, Sonder suggested the existence of a network of primary faults in the Earth's crust, manifested as "lineaments" - rectilinear structures and landforms.

Г. Kloos and R. Staub believed that the structure of Western Europe could be better understood by assuming that the Earth's crust is divided by deep rifts into blocks, each of which moves as a single unit.

"Geometric regularity of location of deep tectonic faults (V.V. Belousov), indicates that deep processes underlying vertical movements of the Earth's crust, developed in the space of the subsoil not disorderly, but along some lines, mostly straight and subordinated to certain directions. Even when, at first sight, the zones of uplift and deflections seem to form smoothly curved arcs (Carpathians, Verkhoyanskiy Range, Western Alps), a closer examination shows that such arcs consist of separate rectilinear segments with a variation of strike-slope at some angle" [5].

"Simultaneous manifestation (according to V.V. Belousov, 1975), on the

surface of continents of different endogenous regimes, "indicates the heterogeneity of the Earth thermal field: at the same time heat flows in different places are different in their intensity, therefore, heat flows change their intensity both in space and in time" [9]. This fact, indicates the existence of a single control mechanism, under the influence of which the system and objects, in its geological space, evolve.

The geometric regularity, discreteness, and periodicity of the location of zones of tectonic fault systems, indicates the symmetry of the Earth system.

This circumstance, makes it possible to widely apply the method of analogy in geology.

Структура разлома (по В.А. Невскому, 1959)

, for the purpose of
mineral deposit
prediction

When analyzing the spatial location of deep faults their systematicity is established (system of deep fault zones of the Urals, Pamir - meridional faults, latitudinal zone of the Southern Tien Shan and Northern Tien Shan fault systems, etc.). "Faults close to each other for 30-40 km. determine (according to V.E. Hain) the formation and development of geosynclines"[5].

The block structure of the Earth's crust is manifested at the lowest level of the hierarchy.

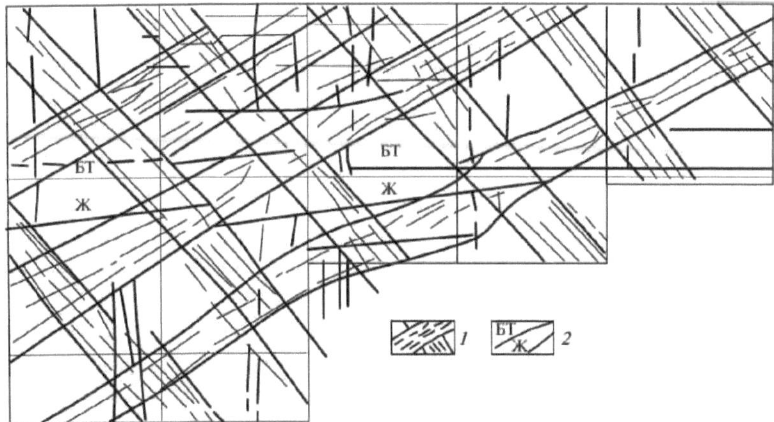

Schematic. Scale: 1:25 000. Analysis of the location of tectonic faults in the Earth's crust. Central Asia, Kuramin Ridge (60 km). The elementary blocks of the Earth's system. Location of paired faults in the Earth's crust. (Geological map by B.D. Lyashkesvich, 1988).
1 - intersection nodes of paired faults; 2 - BT - Bashtavak, G - iron faults [Ustyantsev, 1989]

Map. Regular arrangement of structural elements in the Earth system space. The Kuramin middle massif of block structure. The map is compiled by the method of separate analysis of fault and fracture tectonics (V.N. Ustyantsev, 1988).

"Solid state physics, considers two possible modes of fracture: fracture and shear. Fracture is the nucleation of cracks and their growth. It is established experimentally that in a crack the stress is concentrated at the upper end and increases with the length of the crack. Consequently, for a given stress applied externally, a crack that exceeds its critical length ("Griffiths crack") will spontaneously enlarge. Additionally, note that the tensile strength of many brittle bodies is strongly dependent on the time for which the tensile stress is applied and increases with decreasing load time. When the tensile stress created by the reflected pulse exceeds the tensile strength of the material, fracture will occur. This phenomenon is known as "pitting" or "spalling". Such fractures have been called "Hopkins fractures". The formation of three-dimensional ore-bearing zones determines the mechanism by which fracture and breccia zones are formed. Rocks fracture under forces that promote fracture and shear. Fracture consists in nucleation of cracks and their growth. In a crack, stress concentrates and increases with crack length, i.e. if a given stress is externally applied, a crack that exceeds the critical length will spontaneously enlarge. The fracture of rocks begins where the energy causes the stress field to appear, the potential of which is higher than the strength of the rock.

The tensile strength of rocks is 6-15 times less than their compressive strength, i.e., destruction begins in the tensile regions. In the tectonosphere, where lithostatic pressure is always present, relative compression and tension are possible. If one of the three orthogonal vectors of compression is the smallest, it defines the field of relative tension, i.e. rock destruction in tectonosphere conditions begins where the potential of relative tension exceeds the limit of resistance of specific rocks to tension. These conditions are possible under resonance of even low-amplitude waves, and, therefore, three-dimensional fracture-fracture areal can be considered as a field of interaction of coherent wave flows arising from reflection in any block of the tectonosphere". (V.V. Bogatsky, 1986) [10].

1. the longer the wavelength, the larger the areas into which the waves penetrate;

2. The beating effect reflects the occurrence of alternating resonant compression and extension zones even at the superposition of waves of different lengths and also explains the periodicity and discreteness of the spatial manifestation (laterally and vertically) of the compression (schist, buckling) and extension (detachment, brecciation) zones, in which case a compression-strain wave (a longitudinal wave from a point source) occurs;

3. the amplitude along the wave front always decreases gradually, with a noticeable decrease only at distances comparable to the wavelength; the longer the wavelength, the larger the areas to which the waves penetrate;

4. When the resonance energy is higher than the fracture resistance of the medium, fracture-breccia or weakened zones arise. If the stratified section has a number of reflecting surfaces associated with different physical properties of adjacent strata, in addition to the daytime, the uppermost reflecting surface, systems of intermediate reflecting screens will appear, which will cause the appearance of a series of compression or tension surfaces, and as a consequence, the formation of a series of flat fracture-breccia zones of fissility and compression.

If a wave-beam passes a certain interval, then on all its crests, corresponding to the beams, impulses are generated, generating conditions in which extension zones and compression zones, subparallel to the Earth surface and interface surfaces of the formations composing the section, systems of parallel "consonant" with the bedding of sedimentary-volcanogenic strata, or their oblique fracture-breccia zones and delamination zones are formed. Physico-chemical deformations are genetically connected with stress waves and fields, which are determined by the hierarchy of tectonic disturbance systems.

Mineral raw materials (of any type) are confined to intensely dislocated, screened strata - zones of compression (islands), and within them - to local areas of tension (fracture-breccia structures). In this case, repeated change of compression conditions by tensile conditions can lead to high concentrations of noble metal and other minerals.

That is, a wave mechanism for the concentration of a useful component is defined, the genesis of which is linked to stationary energy centres that are genetically linked to the Earth's auto-oscillation system.

Analysis of the localisation conditions of minerals, indicates their association with zones of increased permeability regardless of the composition of the host rocks. An important feature is the combination of two or even three mutually orthogonal structural forms of intense permeability. They can be sub-vertical, oval cross-section, cylindrical channels, linear zones, as well as sub-horizontal and plate-like bodies, which have a fracture-brack structure. Migration of hydrotherms and fluids takes place along the flat fracture-breccia zones. This mechanism explains the formation of sills, which lie unconformably across beds and sections. The flat-lying and steeply lying fracture-breccia zones predetermine conditions of localization and migration of a substance from deep horizons to overlying ones (such migration processes are currently fixed in the stretching zones - potential accumulators of hydrocarbon raw materials in the fins" [[5,7]). [[5,7].

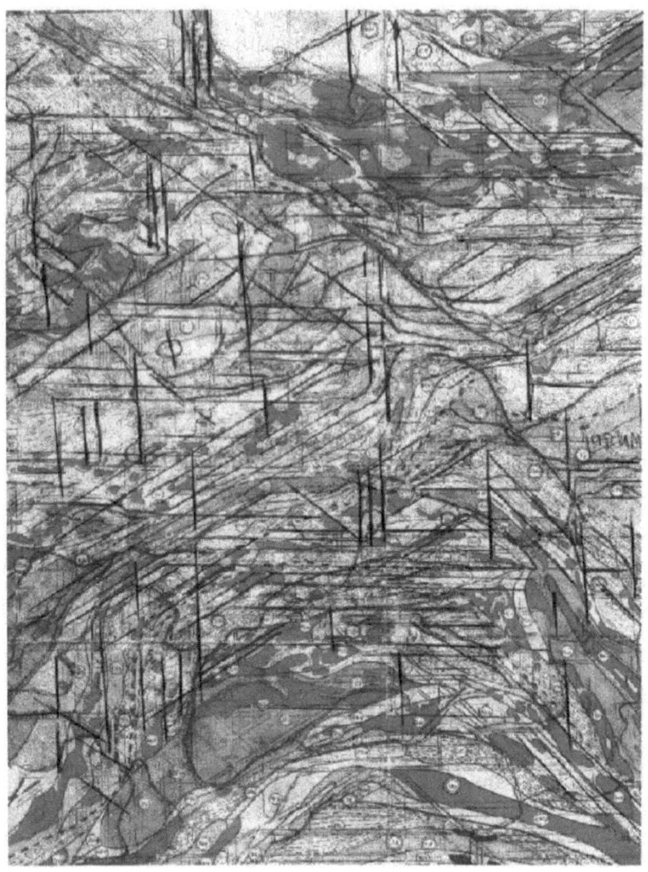

Fragment of the map by V.I. Popov, - structural-formation zoning of eastern Central Asia. M. 1:2 000 000.

Regular arrangement of structural elements in the space of the Earth system. (V.N. Ustyantsev, 2007).

The junction of the Pamir, Tarim, Southern, Middle, Northern Tien Shan and Karakoram structures. The junction is controlled by faults. Legend: blue - sedimentary formations; red -

granitoid formations; green - basite formations; black - ultrabasite formations.

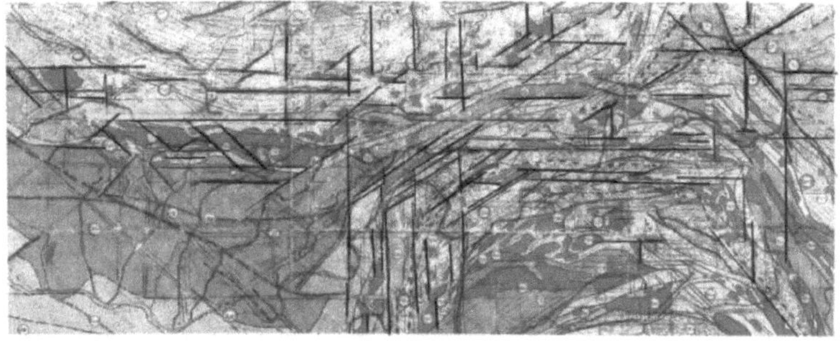

Map. Regular arrangement of structural elements in the space of the Earth system. Junction of the Southern Tien Shan, Northern and Western Pamirs and Kucn-Lun structures. Ustyantsev V.N., (2007) In the south-east - Tarim.

Map. End junction of the structures of the Middle (west) and Northern Tien Shan along the Talas-Fergana fault. The Fergana Depression in the south-west, the Naryn Depression in the north-east, and the Tarim Craton in the south-east. Structure

Legend:

blue - sedimentary formations;

red - granitoid formations; green - basite formations;

black - ultrabasite formations.

Map. Gissar, Karakum-Tajik middle massifs (southern part), in the north - cordillera structures of Southern Tien-Shan. South-east - junction of north-western structures of Pamir, South Tien Shan Gissar core and Karakum-Tajik middle massif transformed into trough.

Marginal faults. The great importance of marginal faults in the history of crustal development was pointed out by V.A. Obruchev and V.I. Popov (1938). V.I. Popov called marginal faults "discordant lines", and believed that they are large faults syngenetic with sedimentation formation, which divide areas of consonant and non-concordant accumulation of deposits (usually separated in both areas by thickness and facies composition). This avoids the assumption of tectonic convergence of facies, which is unlikely with the sustained steep dip of the separating faults. He also noted the marginal position of the faults in relation to the strike of the main structures.

The analysis of marginal faults has shown that they are a group of faults, longitudinal (consonant) in relation to the strike of geo-anticlinal folded structures - zones of increased deformation of the earth's crust, it is closely related to their development. At the same time, marginal faults are composed of separate segments of regional faults of different strike lengths. A common feature of marginal faults are disjunctive boundary dislocations that separate different structural forms in sign, peculiar boundaries of changing thicknesses and sediment types of characteristic ore occurrences and magmatism. This system of steeply dipping faults, accompanied by zones of crushing, fracturing and increased metamorphism, is often accompanied by belts of various types of mineralisation. Marginal faults limit ancient platforms and their activated protrusions from geosynclinal belts: the Donbass-Ural, Donbass-South Tien Shan and Central Asian.

The interest in the median massifs was stimulated by the fact that they are characterised by a variety of rich deposits. The Kuramin massif is characterised by complex ore formations: skarn-polymetallic, copper-porphyritic, quartz-silver-sulphide, quartz-copper-bismuth, gold-sulphide, gold-antimony, skarn-magnetite, skarn-molybdenite-shelite. There are also low-temperature (silver) - lead-zinc, barite-carbonate-fluorite, alunite and other formations. General geological studies have shown that in the zone of the forty-second parallel, there are the largest deposits of various types of minerals, including HC and diamonds.

The median massifs, marking zones of crustal extension, are the "key" to understanding the formation pattern of the Earth system and its minerals (minerals of all types, from Archean to Quaternary). In each block of the median massif, 1 to 3 large deposits of solid minerals and HC are localised in the adjacent negative geoform. Positive and negative geoforms, controlled by zones of deep fault systems. HC are localized in negative geoforms: Structural traps are formed in the post-orogenic period - the era of denudation processes (the process of degassing of the Earth system, occurs continuously, but with varying degrees of intensity). Permeable zones of the Earth's crust are marked by areas of development of median massifs and areas of development of postgeosynclinal massifs of granitoids (both of them mark the zones of crustal expansion). An important role in the process of directional migration of hydrocarbons and hydrotherms, play the processes of granitization and basification of the mid- massifs (screening effect), according to which, hydrocarbons and acidic thermals, migrate to the direction of least pressure along the marginal fractures and are localized in the negative geoforms (structural traps) and the basement. The formation of ophiolite belts, permeable zones of deep fault systems, is associated with the rotation of the Earth and zone melting processes both toward the core and toward the crust. Degassing of fluid-saturated magma is a source of HC. Injection of such magma occurred during eras of crustal destruction, the acting factor being the occurrence of overpressure on the part of mania.

The formation of energy barriers is a vivid manifestation of the Earth system ***energy.*** Under the influence of system properties of EEC energy waves of the first and second kind, the area of generation, migration and localization of mineralogical associations of various types and kinds is formed. Under the influence of energy waves, the mechanism of ore bodies formation and wave mechanism of useful component concentration in the Earth's crust blocks is working.

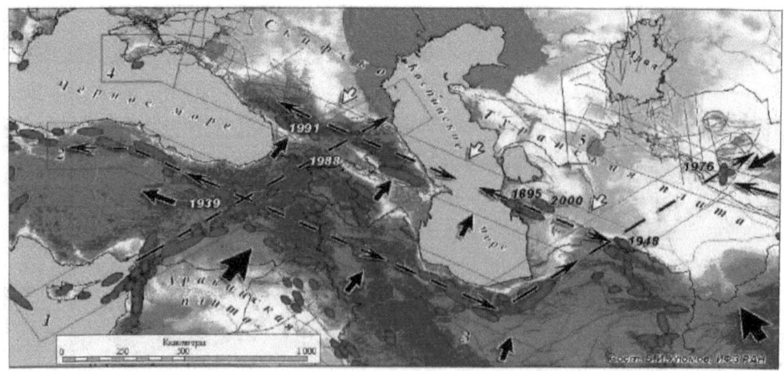

The primary ERS map.

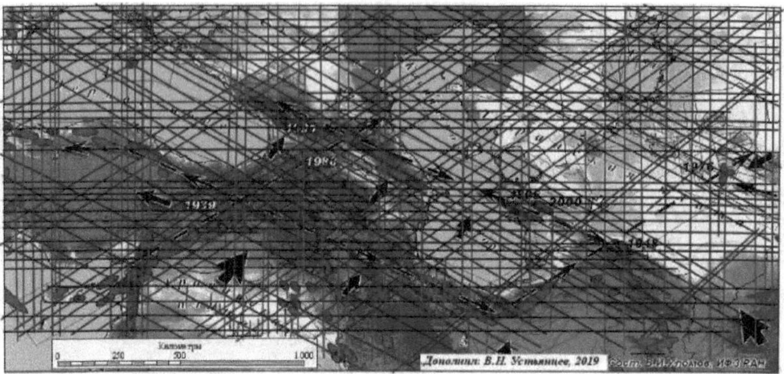

Map. The regular arrangement of the structural elements in the space of the Earth system. Analytical structural map. Crustal and mantle objects, hydrocarbon deposits, earthquake faults - controlled by faults (straight black and red lines - faults of discontinuous nature; blue - shear; brown - thrusts). There is a wide band from the Carpathian Mountains to Western Siberia - a zone of deep fault systems of meridional strike, which is global in scope and controls deposits of hydrocarbons, gold and other minerals. There are nine to ten zones of fracture systems in this belt. The zones of latitudinal strike-slip fault systems are distinguished: the black lines are faults of discontinuous type, and hydrocarbon outcrops are faults of discontinuous type. Zones of NE strike-slip fault zones (about five systems) - thrust faults, - structural traps, - brown lines.Zones of shear disruption zones (about five systems), - blue lines are identified. **Zones of tectonic disturbance systems girdle the Earth system.**

A matrix of Earth system objects. Geometricisation. Main four strike directions of tectonic structures. Moving belts. Tectonic faults and movable belts. Compiled by - V.N. Ustyantsev, 31.07. 2019.

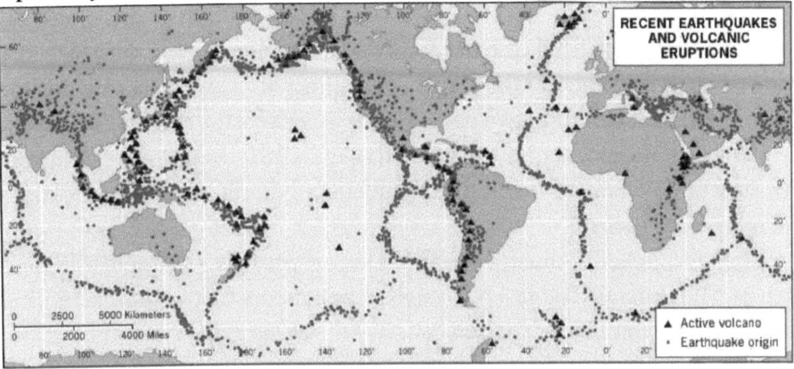

Map of volcanoes and earthquake zones.

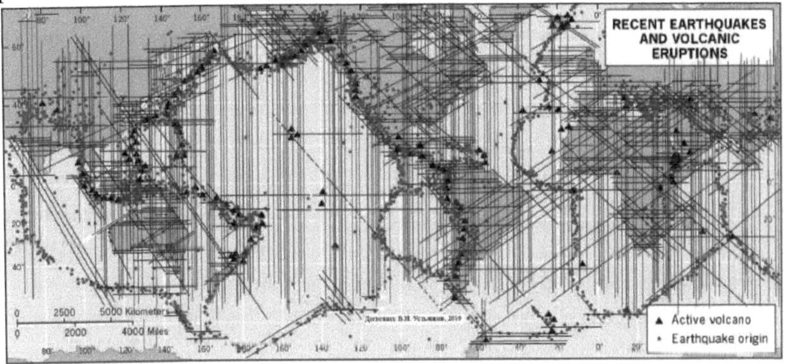

Map. Regular arrangement of structural elements in the Earth system space. Zones of tectonic disturbance systems in four main directions are distinguished: black and red lines are tectonic disturbances of discontinuity type; blue - shear, brown - thrusts. The control of Earth system objects by zones of tectonic disturbance systems is clearly shown. The main earthquake zones are marked by zones of deep fault systems, the four main directions, and tectonic faults control HC deposits. Compiled by: V.N. Ustyantsev, 21. 09. 2019.

The Earth's *main latitudinal structure [by A. Anokhin] [5]* is the equatorial zone of linear dislocations along which the *leftward shift of the northern hemisphere relative to the southern hemisphere* develops. To the north and south of the equator, latitudinal "critical" belts alternate about 20° apart.

The main "meridional" line appears to be the Earth's rotational axis, so on the surface it is expressed by a number of linear structures of the 2nd order -

submeridional lineaments alternating through 20°, 40°, 60°, 90°, which include a number of land and oceanic bottom ridges, fragments of mid-ocean ridges, island arcs.

The largest diagonals for the Earth are most likely two diagonal planes passing through the centre of the planet and inclined to its rotation axis at about a 45° angle. These planes form two circles as they intersect the surface of the planet:

- chain of lineaments SE edge of Asia - main NE - Indian Ocean lineament - NW structure of South America - Cordilleras in North America - closing the circle in the Bering Sea.

Lineament chain Sumatra - South Asia - Caucasus - Tornquist line - NE branch of Mid-Atlantic Ridge - NW edge of South America - East Pacific Rise - circle closure south of New Zealand.

On a relief map of the world, these diagonal lineament zones can be seen as bi-period sinusoids composed of heterogeneous linear shapes. Diagonals of the 2nd order can also be distinguished without much difficulty (e.g. subparallel NW chains of islands in the central Pacific Ocean, the Red Sea-Apennines line, etc.), they also alternate with an approximate step of 20°.

At all depths, the internal structural lines of the Benioff zones have a similar orientation, generally corresponding to the directionality of the global discontinuity network (especially in the orthogonal part). The resulting rose-diagram of the directivity of the linear structure elements of all Benioff zones shows an even better correspondence. Along with other results, the configuration of these rose diagrams leads to the following conclusions:

-the regularities of the structural plan of the seismic-focal zone of the Pacific movable belt correspond to the regularities of the global structural plan of the Earth, determined, in turn, by the pattern of the regular planetary lineament-disjunctive (discontinuous) network.

-the general orientation of the linear structures of the seismic focal zone is subject to the azimuthal patterns of the global rupture network of the Earth throughout the depth of manifestation of the seismic focal zone, which determines the depth of this network to the base of the tectonosphere.

The results of the directional rose diagrams, as well as the results of factor analysis, support the rotational nature of the stress network and the global discontinuity it generates:

- the orientation of the global network is symmetrical about the planet's axis of rotation;

- "crowding" of factor loads at the equator is a sign of centrifugal forces;

- The asymmetry and reversal of the sign of the factor loads when crossing the equator is a sign of the Coriolis force;

- The pulsating nature of the manifestation of the main factors is a sign of periodic changes in the planet's rate of rotation" [Anokhin].

Hierarchy of tectonic faults forms a "rigid controlling system of geological processes" of the Earth's tectonosphere. The migration of faults occurs in the direction opposite to the direction of the Earth's rotation and this is obviously a pattern that reflects the properties of the Earth's envelope region and also indicates the genetic relationship of tectonic faults and folded structures with the upper mantle.

Faults control and guide the movement of magma. Faults have three types of plume fractures (outcropping at an acute angle, at right angles and "torn off", sub-parallel to the fault). Plumage fractures often contain ore bodies. The ability of faults to reflect and conduct elastic waves is reflected in the formation of fracture-brack zones along their strike, the genesis of which is associated with wave effects (resonance and interference). The block structure of the Earth's crust contributes to the control over the migration processes of matter, in the zones of their influence. Deep faults control not only the placement of igneous formations but also sedimentary ones, which is well reflected in the formation-structure map by V.I. Popov. Sedimentary formations are often hydrocarbon accumulators.

ARCTIC CRUST THICKNESS MAP

C. N. Kashubin, O. V. Petrov, E. D. Milstein, E. A. Androsov, A. F. Morozov,
V. D. Kaminsky, V.A. Poselov

The crustal thickness map is derived from deep seismic surveys and gravity field anomalies in the Circumpolar Arctic. More than 300 profiles with a total length of about 140,000 km and correlation equations linking the depth of the Moho boundary with Bouguer anomalies and topography were used to construct the map. The digital model of the Circumpolar Arctic crustal thickness map constructed from these data is represented by a 10*10 km grid.

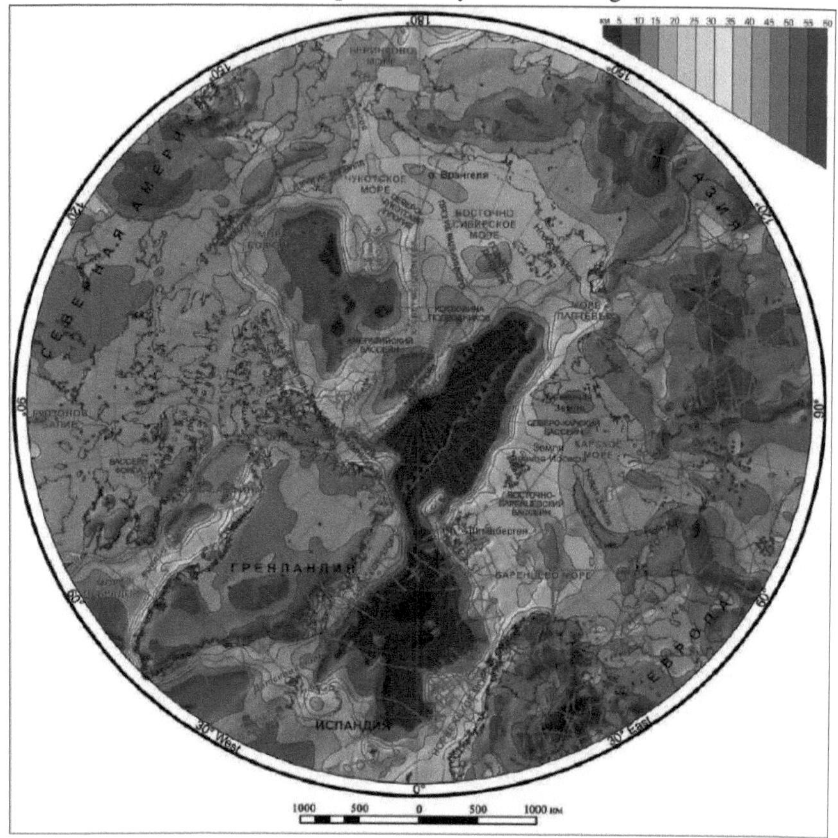

Figure 14: Crustal thickness map of the Circumpolar Arctic [Kashubin et al., 2011; Kashubin et al., 2014].

The grey lines show the main depth seismic profiles, and the grey dots show the seismic stations whose materials were used in the construction of the map.

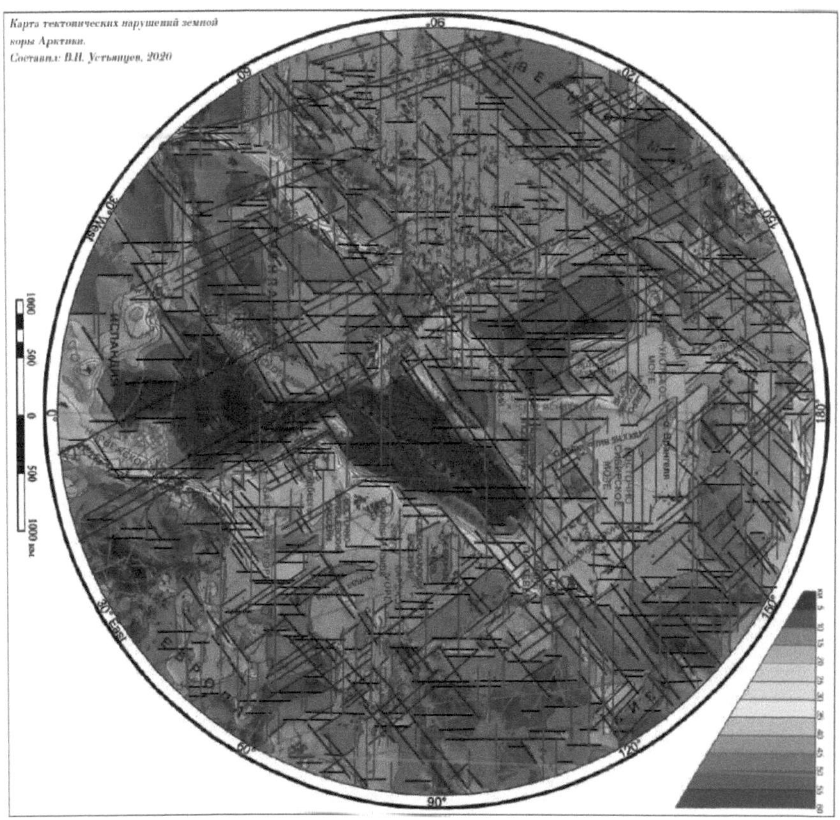

Map of the tectonic fault system zones of the Circumpolar Arctic region. The regular arrangement of structural elements in the Earth system space is controlled by tectonic disturbances of four directions. The system formation has taken place: vaulted uplift - oceanic depression. The zones of intense permeability (deformation) are associated with the formation of large and giant mineral deposits of various types.

Due to the fact that:

"The system must be open. A closed system, in accordance with the laws of thermodynamics, must eventually reach a state with maximum entropy and stop any evolution" (I. Prigogine). That is to say, the process of mineral formation is anti-entropic. The open system is formed thanks to the hierarchy of tectonic disturbances. Thus, the zones of tectonic disturbance systems are the main factor under the influence of which mineral raw material deposits are formed.

Compiled by: V.N. Ustyantsev, 2020.

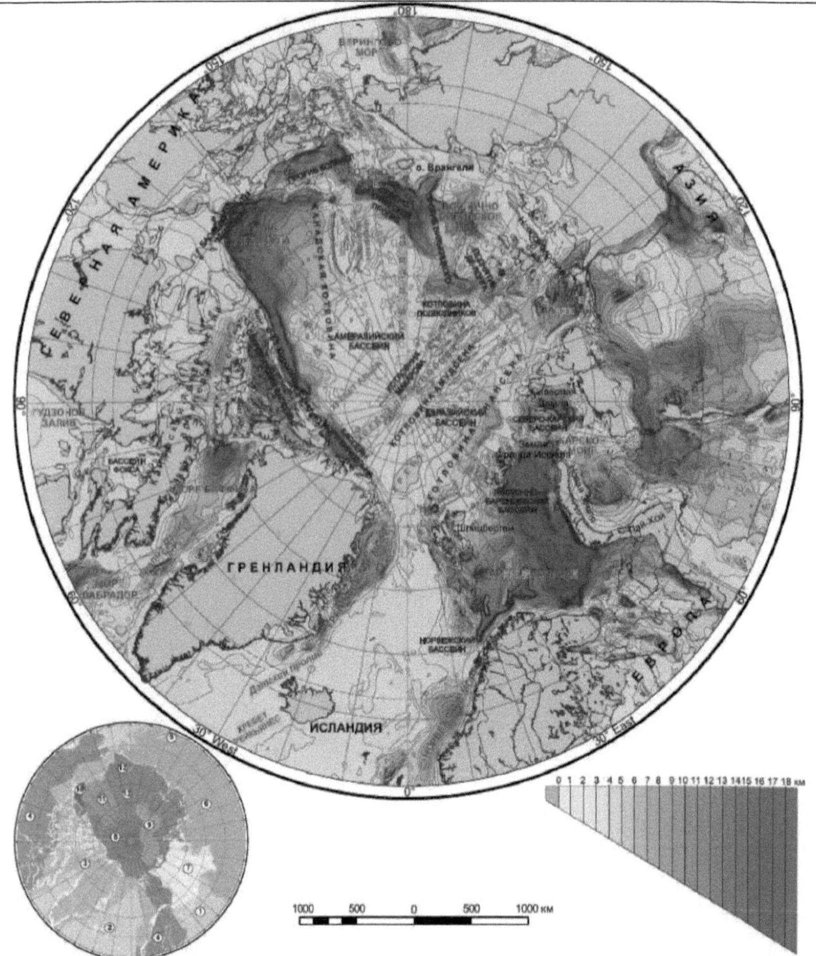

Figure 18: Map of the thickness of the sedimentary cover of the Circumpolar Arctic [Petrov et al., 2016].

Author's schematic: 1 - Yu. M. Erinchek et al., 2002 (stock materials). Relief map of different-age basement of the East European Platform and Timan-Pshora province; 2 - Divins D. L., 2003. NGDC Total Sediment Thickness of the World's Oceans and Marginal Seas; *3* - Grantz A. et al., 2010. Map showing the sedimentary successions of the Arctic Region that may be prospective for hydrocarbons; *4* - Laskc G. and Masters G. A., 2010. Global Digital map of Sediment Thickness; 5 - T. C. S. Sakulina et al., 2011. Sedimentary basins of the Okhotomor region; 6 - S. P. Shokalskiy et al., 2010 (stock footage). Schematic thickness map of the sedimentary cover of the Urals, Siberia and Far East; *7* - T. S. Sakulina et al., 2011. Map of the thickness of the sedimentary cover of the Barsntsevo-Kara region; *8* - V.A. Posslov et al., 2012. Map of the thickness of the sedimentary cover of the Arctic Ocean; *9* - K.G. Stavrov et al., 2011. Map of the sedimentary cover thickness at a scale of 1:5 000 000; *10* - N. Kumar et al., 2010. Tectonic and Stratigraphic Interpretation of a New Regional Deep-seismic Reflection Survey offshore Banks Island; *11* - D. C. Mosher et al., 2012. Sediment Distribution in Canada Basin; *12* - H. A. Petrovskaya et al., 2008. Main features of the geological structure of the Russian sector of the Chukchi Sea; *13* - I.Yu. Vinokurov et al. 2013. Sedimentary cover thickness according to the results of seismic profiling of the Arktika 2012 expedition.

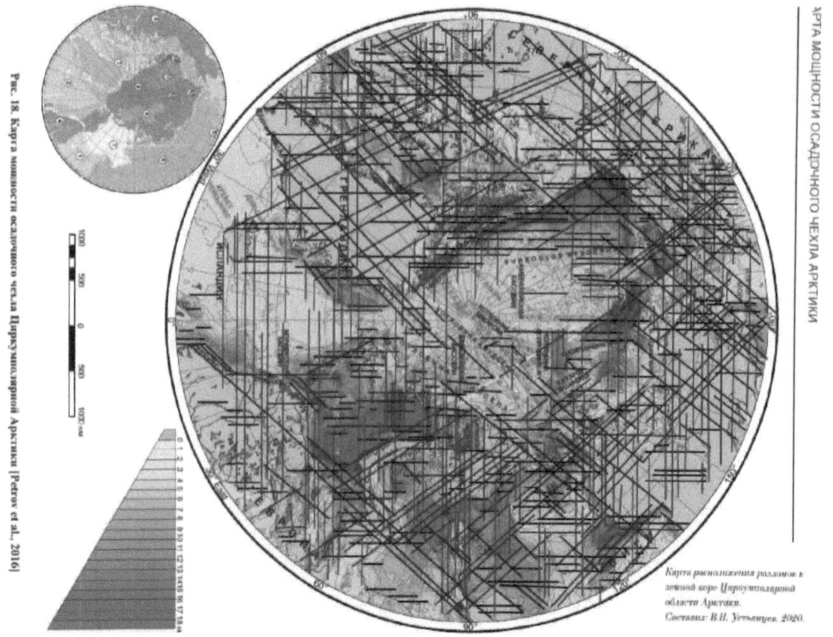

Map. Regular placement of faults in the Earth's crust. There are western, central and eastern extensive zones of meridional fault systems. The zones of the fracture systems control the location of crustal structural elements and mineral deposits. Zones where the thickness of sedimentary formations reaches 13-18 km - zones of crustal stretching are fixed. Such zones are favorable for the localization of hydrocarbons. The intersections of both individual faults and intersections of fault systems are also favourable for localisation of minerals.

Geological map of India

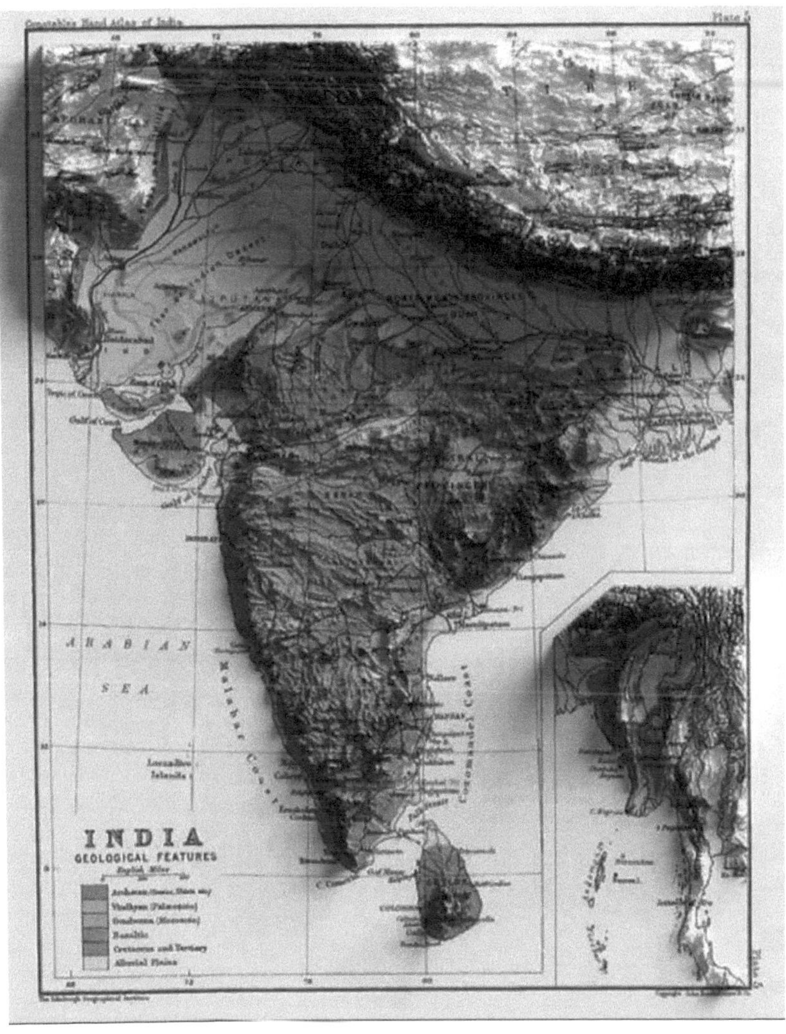

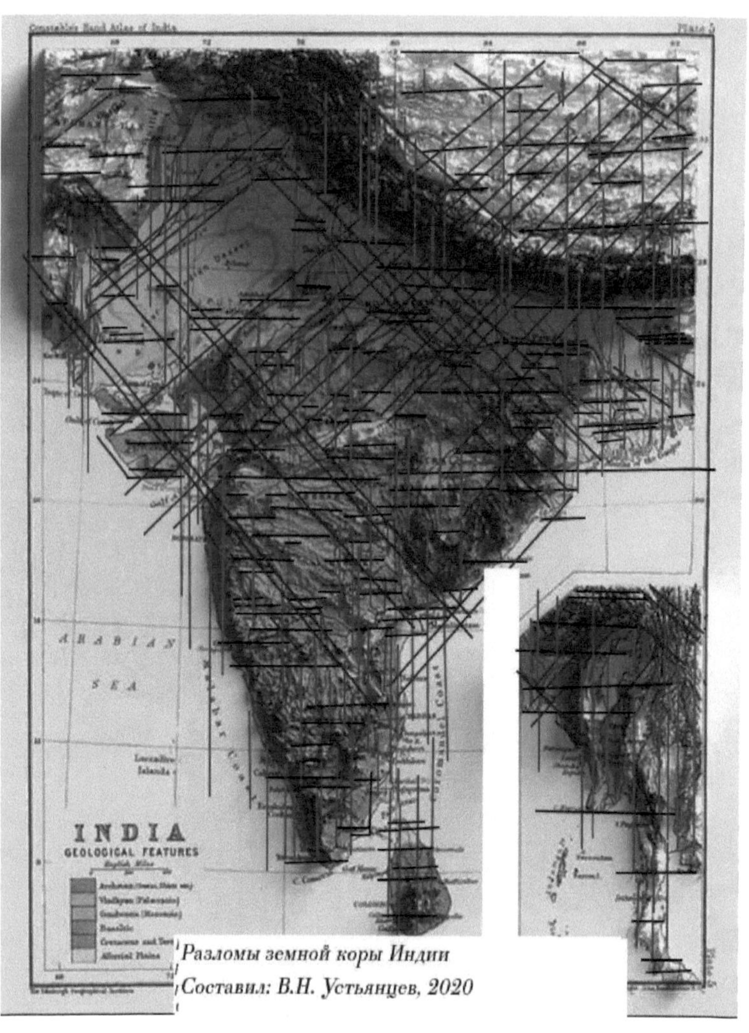

Разломы земной коры Индии
Составил: В.Н. Устьянцев, 2020

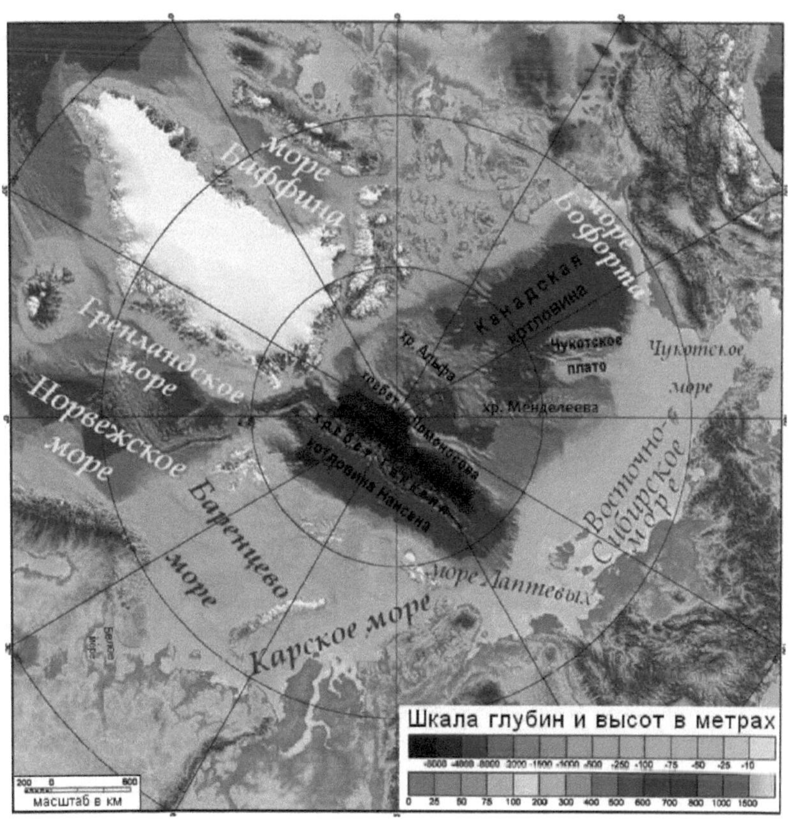

море Баффина
море Бофорта
Канадская котловина
хр. Альфа
Чукотское плато
Чукотское море
хр. Менделеева
хребет Ломоносова
хребет Гаккеля
котловина Нансена
Восточно-Сибирское море
Гренландское море
Норвежское море
Баренцево море
Белое море
Карское море
море Лаптевых
Шкала глубин и высот в метрах
-6000 -4000 -3000 -2000 -1500 -1000 -500 -250 -100 -75 -50 -25 -10
0 25 50 75 100 200 300 400 500 600 700 800 1000 1500
200 0 600
масштаб в км

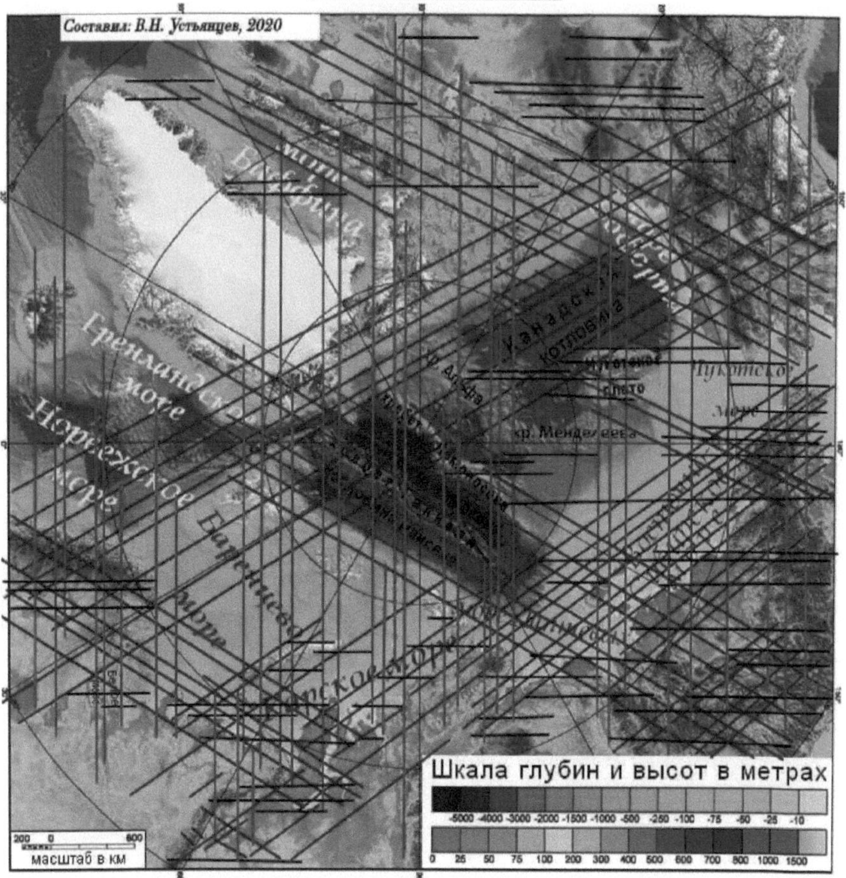

A map of the location of tectonic fault systems in the crust of the Arctic region. A weakened area of the continent has formed under the influence of the tectonic disturbance system zones of the four main directions, which is the area of localization of seas. Tectonic disturbances controlling the depressions and ridges separating them are fixed on the continent. In northeastern Russia, a junction where tectonic fault systems intersect has been identified. This intersection is very favourable for the localisation of minerals.

Promising for the localization of mineral resources are the intersections of zones of tectonic fault systems and the zones themselves, both on the continent and in the adjacent seas. The map reflects the mechanism of the formation of the continental marginal seas and the Mediterranean seas. The map provides clear evidence that the method of mapping tectonic faults is correct.

Systems of tectonic faults controlling oceanic depressions determine the mechanism of formation of the continental margin seas, which indicates the time of their deposition (Archean). Given the existence of slow oscillatory movements of the lithosphere (epeirogenic oscillations), the mechanism of formation of the system: vaulted uplift - oceanic depression is determined. Compiled by: V.N. Ustyantsev, 2020.

Examples:

Oil and gas bearing formations of the Arctic and other regions of the world:

"Ancient NGMT AUVs of the Paleoproterozoic, Riphean and Early Vendian

(~ 2,100 - 570 Ma). The isolation of the Paleoproterozoic AUF NGMT is confirmed by the existence of metamorphosed black shale formations of Early Proterozoic (Paleoproterozoic) age on many Precambrian platforms.

Early Proterozoic (paleoproterozoic) NGMTs in Russia are found on the Taimyr Peninsula north of latitude 76o N, where they are represented by two terrigenous strata: the Octozoic (unconformably deposited on the Archean metamorphic basement, 2,500 m thick) and the Zhdanovka (conformably deposited on the Octozoic strata, over 3,200 m thick).

Universitetskaya (C2) - Severnaya Zemlya Arch., Kuonamskaya (C1-2) - Yakutia, Gravinorechenskaya (C1-2) - Northern Taimyr, Inikanskaya (C1-2) - Yakutia, Shuminskaya (C1) - Tungusskaya syneclise, Maratovskaya (C1) - Arch. Severnaya Zemlya, Akrinskaya (C1-2) -Yakutia, Awatage (C2) -West China, Xiaoerbulake (C1), Lower Arthur Creek Shale (C2) - Australia. The age range is three stages of the Cambrian system: Lower Cambrian - Bothonian and Toyonic, Middle Cambrian - Amgian. The Hanson Glacier Formation (C1-2) spreads in the Franklin Basin, sublatitudinally enveloping North Greenland, is composed of argillites with the Sorg content of 2.5-5.0 %. Residual hydrogen index of these rocks is 400 mg HC/g Sorg. Thickness is 20-40 m. The Sukhan, Chopkotin and Firestone formations (Cz) are distributed in the Sukhan Depression, at the northeastern end of the Siberian Platform. The limestones and clayey limestones are grey and mottled with interlayers and piles of dark-coloured marls, mudstones and schists. The total thickness of these three formations is 500 - 800 m, with an average Sorg content of 0.3-0.5% to 10-17%.

Late Cambrian NGMT AUV (500-480 Ma)

It includes 2 UF HMT and 9 individual HMT: Qiulitage (N) -West China, Ajram (N1) - Jordan, Middle-Upper Cambrian (C2-c) - Europe, Oyuyakhinskaya (N) - Russia, Pay-Khoi, Kurchavinskaya (N) - Arch. Northern Land, Sukhanskaya, Chopkotinskaya and Firenierskaya (N) -North-East of the Siberian

Platform, Verkhneulakhskaya (N) -Yakutia.

Early to Middle Ordovician NGMT AUV (480-460 Ma)

The UF NGMT and 8 individual NGMTs are referred to: Jurassic (O1), Prikolymsky Ordovician (O2), Harkindzhinskaya (O2z) - Kolymo-Omolonsky Massif, Astronomical (O1-2) and Krivolutskaya (O2) of Northern Taimyr, Arch. Taimyr and arch N. Land, Hetuao (O1) -West China, Goldwyer (O21) N.W. Australia, Talbeityvis Formation (O21) - Russia, Pai Khoi.

Silurian NGMT AUV (445 - 420 Ma)

Includes 6 UF NGMTs and 2 individual NGMTs: Sodus Shale (S1) -B. Canada; Pitinga (S) -Brazil; Tannezuft "Hot Shale" (S1 1) -(h sites in Algeria, Tunisia, Libya); Lower Silurian (S1 1) -Morocco, Mauritania, Western Sahara; Fotmigoso (S1), Lower Silurian (S1 1), Silurian (8z ld) -Europe; Lafaiet Bagt (S1), Wolfland (S1) -North Greenland; Akkas (S1), Wolfland (S1) -Europe (S1) - Hungary. Greenland; Akkas (S1), Tanf (S1), Qalibah (S1), Bedinah (S1), Qusaiba Shale (S1), Dadas Shale (S), Batra Member "Hot Shale" (S1), Batra Lower "Hot Shale" (S1), Batra Upper "Hot Shale" (S12) - Arabian Plate; Lower Silurian (S1), Longmaxi - (S1) -South. China; Gaojiabian - (S1) -E. China; Ludlovskaya (S2), Kolvinskaya (S2) -Timano-Pechora Basin; Dvoininskaya (S1 1), Moyerokanskaya (S1), Middendorfskaya (S1-2) -North Taimyr; Umba (S1), Polousnoe (S1), Mautskaya (S1) -Kolym-Omolonsky massive; Putukuneinskaya (S1-2) -East Chukchi.

The vast majority of Silurian NGMTs (27 of z2) were formed during the Early Silurian epoch, and mainly in two regions: the northern African-Arabian platform and the Arctic periphery.

In the Arctic region, Early Silurian NGMTs are also represented by graptolitic shales 20 to 150 m thick, with 2% to 11% Sorg content, residual hydrogen index up to 500 mg HC/g Sorg.

Late Paleozoic-Early Mesozoic NGMT AUV

Late Permian-Triassic NGMT AUV (260-225 Ma)

Presented by 6 UV NGMTs and 57 individual NGMTs: Montney (T), Doig Phosphate (T2) - Canada; Shublice (T) - Alaska; Olenek (T1), Yermakovskaya (T2)

Svalbard; Franz Josef Land Anisian (T2 a), Franz Josef Land Yermakovsky (T2 ld)-Arch. Franz Josef Land); Pur Lower (T 2-3), Pur Upper (T 2-3) -West Siberia; R. Bol. Kegali (T3), Khavakchanskaya (T3), Talbygyra (T3), Ruc. Lednikovy (T3) - Kolyma-Omolon massif; Lower Wuerhe (P2), Changxing (P2), Longtan (P2), Lower Triassic (T), Baijiantan (T), Kelamay (T), Balkonquan (T), Haojiiagou (T2-3), Huangshanjie (T2-3), Karamay (T2-3), Xiaoquangou (T2-3), Upper Triassic" [2].

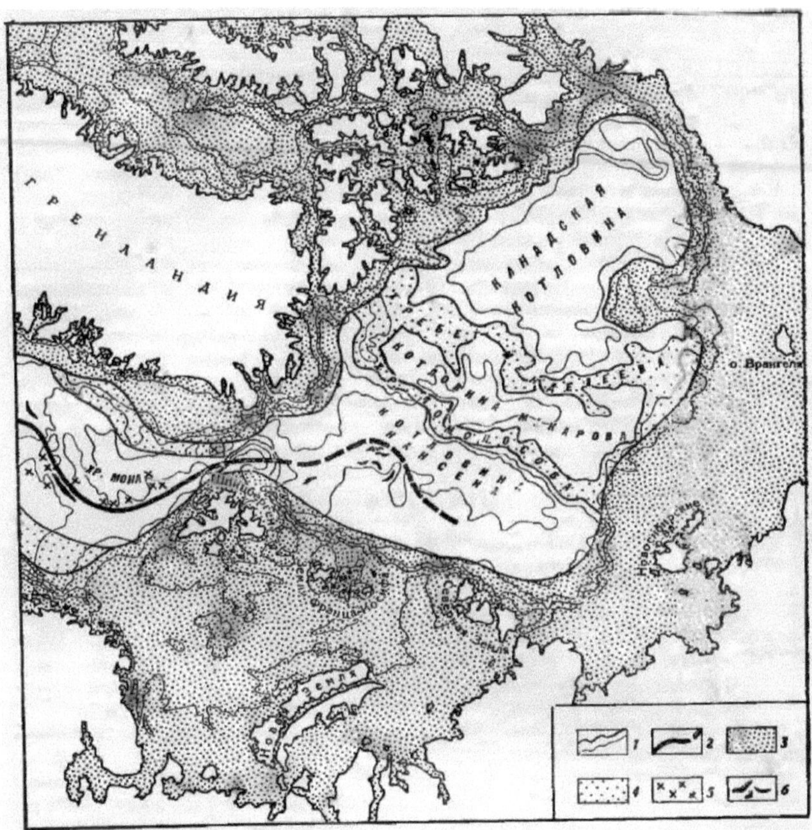

Р и с. 34. Северный Ледовитый океан, схема строения дна

1 — изобаты;
2 — рифтовая зона срединно-океанического хребта;
3 — шельфы;
4 — участки дна с материковым строением земной коры;
5 — подводные вулканы;
6 — отдельные гряды

Шельф этот имеет материковое строение коры и составляет продолжение структур Евразии.

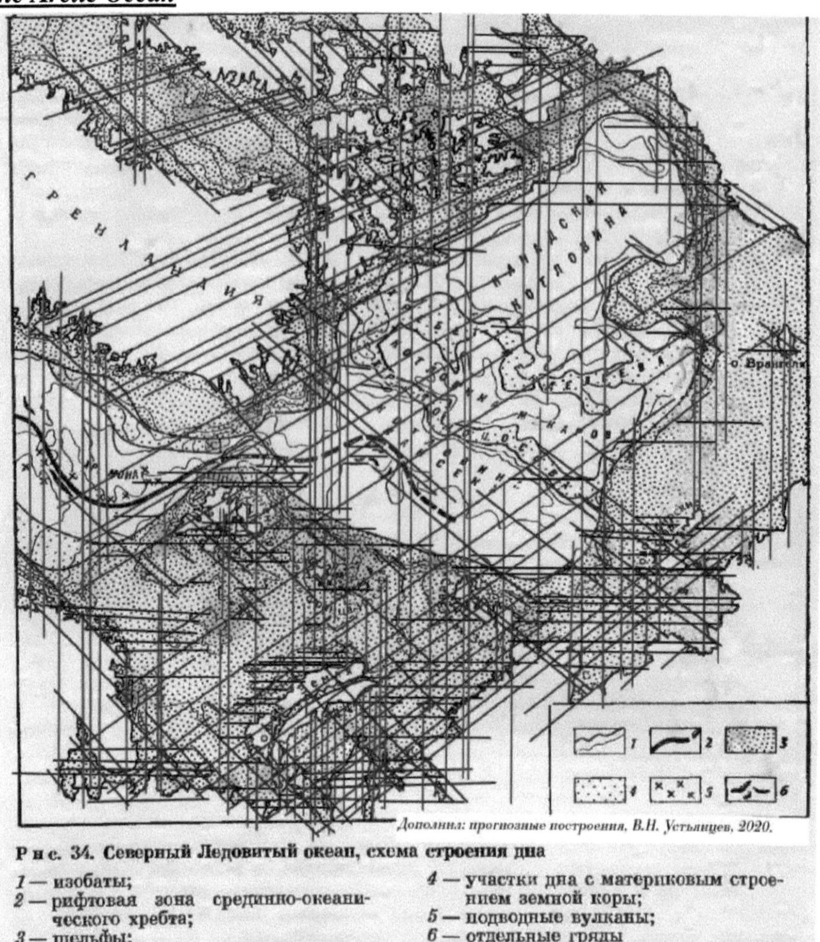

Р и с. 34. Северный Ледовитый океан, схема строения дна

1 — изобаты;
2 — рифтовая зона срединно-океанического хребта;
3 — шельфы;
4 — участки дна с материковым строением земной коры;
5 — подводные вулканы;
6 — отдельные гряды

Detailed map. The strike-slip geometry of the mid-oceanic ridge rift zone structures defines the deformation plan of the oceanic region. The four main strike directions of the structures are clearly shown, indicating that they were laid down in the Archean. The geometry of the rift zone structures reflects the process of evolutionary-directed historical-geological development of the Earth system. The intersection nodes of tectonic fault systems similar to those controlling the White Tiger deposit (Vietnam) are mapped.

Рис. 1. Рельеф дна Северного Ледовитого океана (по Р. М. Деменицкой, 1967)

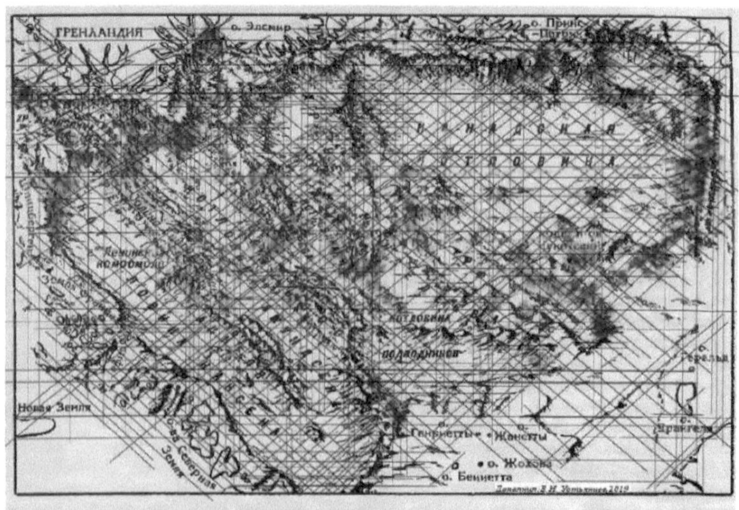

Рис. 1. Рельеф дна Северного Ледовитого океана (по Р. М. Деменицкой, 1967)

A detailed map. Regular arrangement of structural elements in the space of the Earth system. Analytical structural-tectonic prediction. Zones of tectonic fault systems have been identified. Tectonic fault systems that control the geomorphological structures of the Arctic Ocean. Zones of the systems of tectonic faults are located in the Earth's crust, - regularly. The points of intersection of the systems zones, - are the most promising for the localization of mineral deposits, - HC, etc. Compiled by: V.N. Ustyantsev, 25. 09. 2019.

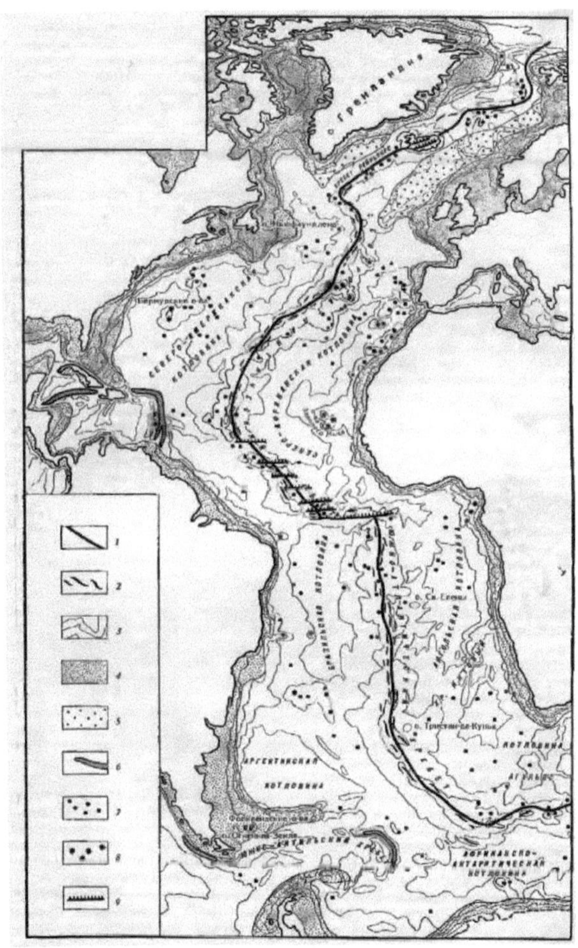

1* and p. 33. Atlantic Ocean, diagram

1 - The rift zone of the mid-oceanic of the mountain range;

2 - individual ridges;

3 - isobaths;

4 - shelf areas;

dpa structures

5 - Areas of the seabed with a continental structure **The earth's crust;**

6 - deep water troughs;

7 - underwater volcanoes;

8 - volcanic islands;

O - main faults

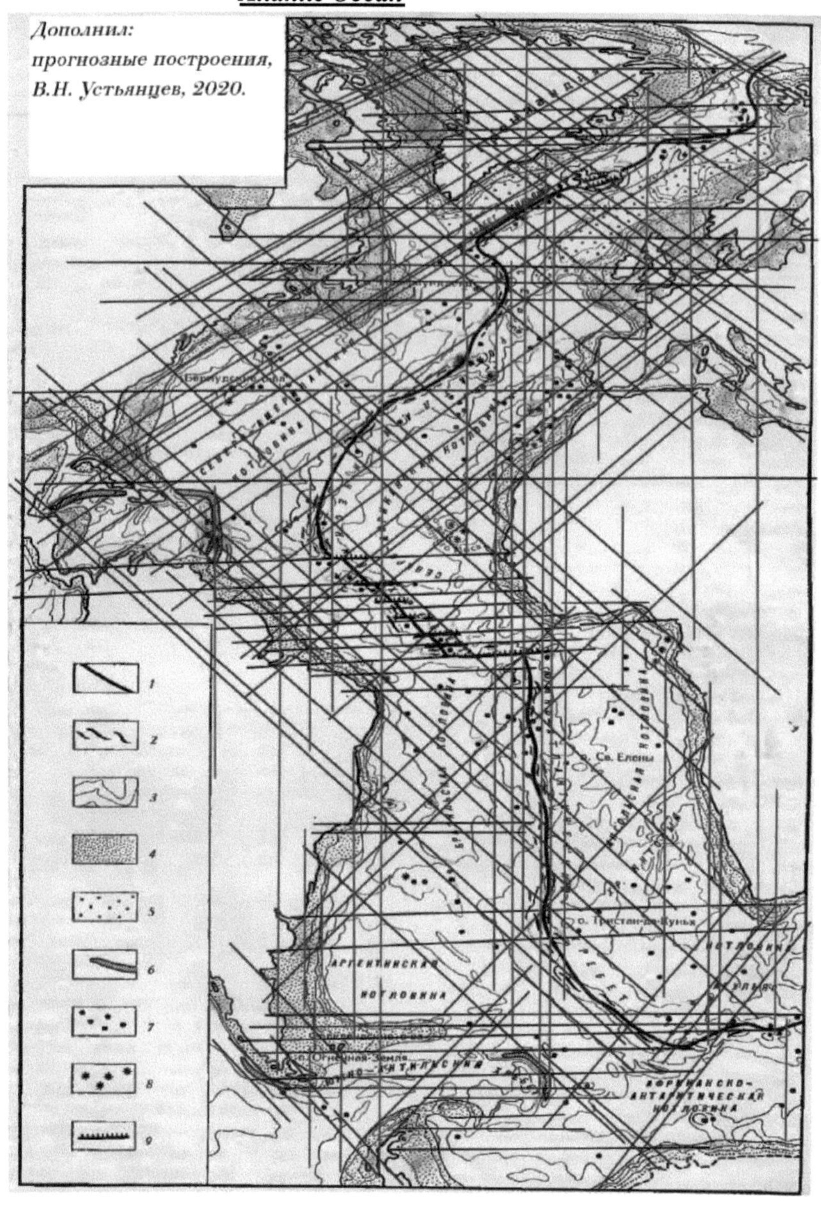

Дополнил:
прогнозные построения,
В.Н. Устьянцев, 2020.
1
2
3
4
5
6
7
8
9

Map. Tectonic fault system zones of the Indian Ocean region

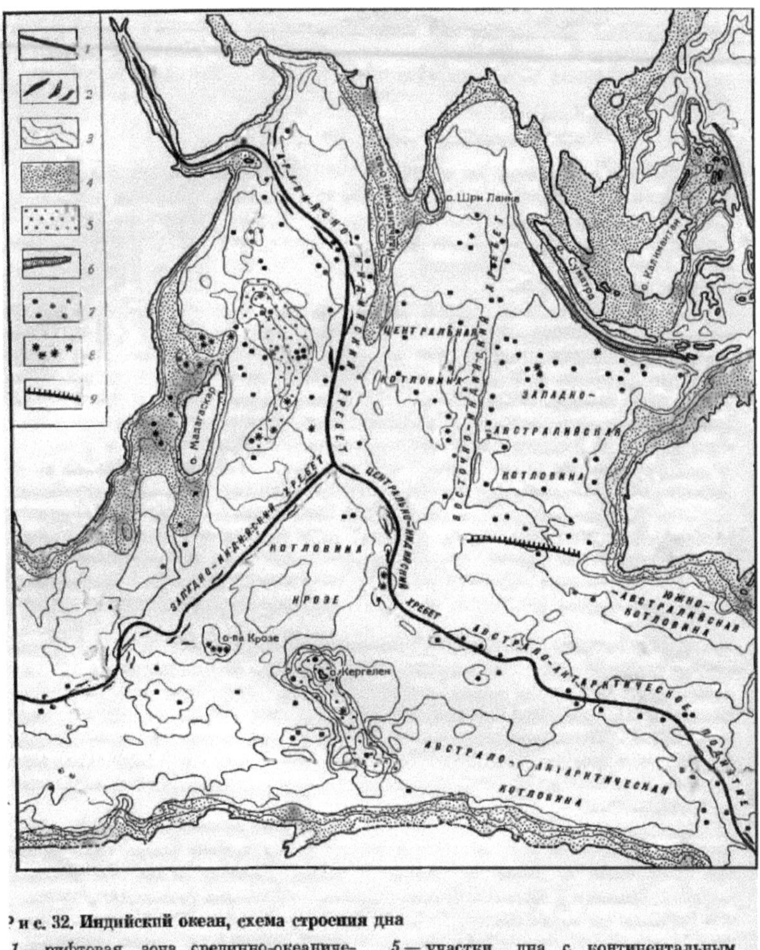

Рис. 32. Индийский океан, схема строения дна

1 — рифтовая зона срединно-океанического хребта;
2 — отдельные гряды;
3 — изобаты;
4 — шельфы;
5 — участки дна с континентальным строением земной коры;
6 — глубоководные желоба;
7 — подводные вулканы;
8 — вулканические острова;
9 — крупнейшие разломы

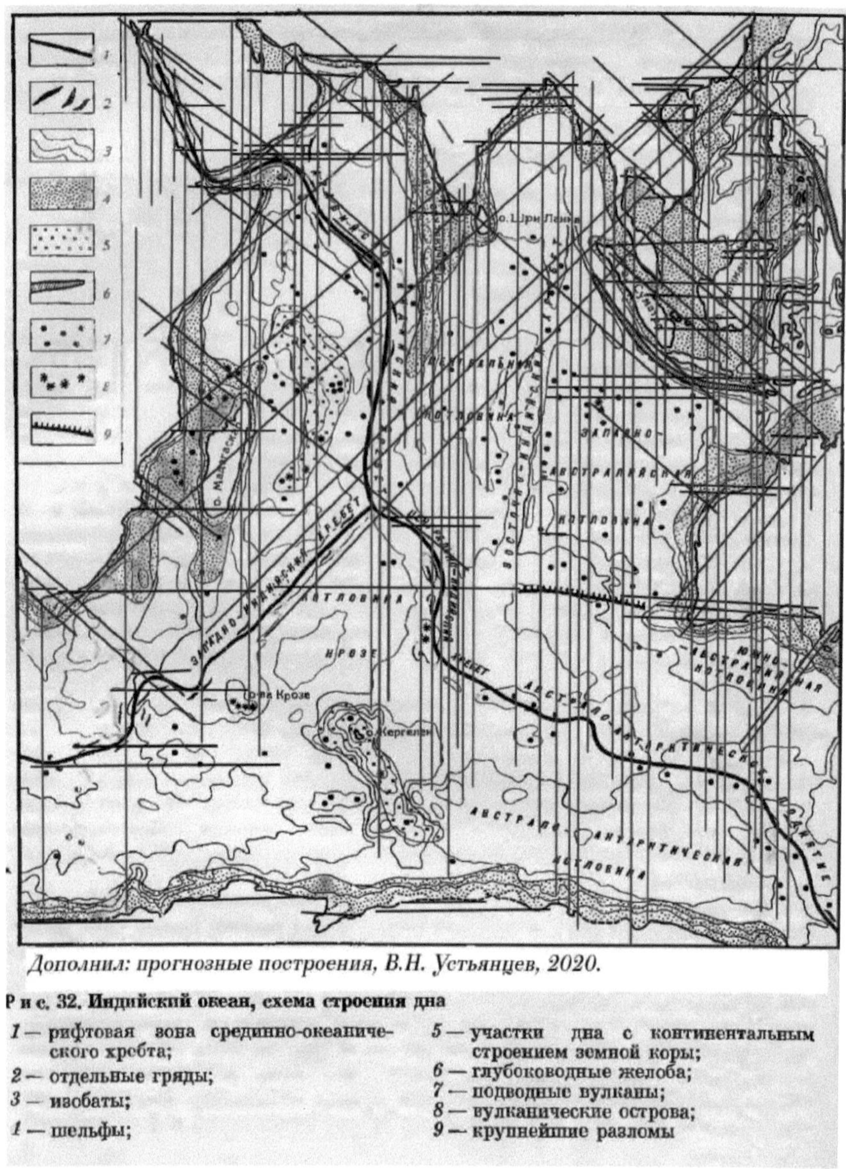

Дополнил: прогнозные построения, В.Н. Устьянцев, 2020.

Р и с. 32. Индийский океан, схема строения дна

1 — рифтовая зона срединно-океанического хребта;
2 — отдельные гряды;
3 — изобаты;
4 — шельфы;
5 — участки дна с континентальным строением земной коры;
6 — глубоководные желоба;
7 — подводные вулканы;
8 — вулканические острова;
9 — крупнейшие разломы

Map. The regular arrangement of the structural elements in the Earth system space. Indian Ocean, - Riftogenesis zones control HC deposits. Compiled by: V.N. Ustyantsev, 2020.

Mid-ocean ridges, as primary structures, determined the deformation plan of the Earth system and the formation of the system: soda uplift - trough.

Researches of H. Gerstenberg, K. Wenzel have shown, that "geochemistry of isotopes of daughter elements of long lived natural radionuclides and especially geochemistry of **Nb and Sr** isotopes, as well as research of oxygen isotope composition in Earth crust have allowed to receive essential results on dynamics and mechanism of substance exchange between crust and mantle, and also on general development of Earth crust. Their results allow us to conclude that:

1. huge continental cores were formed before the turn of 3.0 billion years ago;

2. the growth of continents throughout the Earth's history is related to a sequence of more or less global events, accompanied by high magma activity, which was caused by the uplift of magma from the upper mantle;

3. During the course of the differentiation process, in some areas of the mantle, lithophilic depletion has occurred (in particular - ***rift zones)''*** [15].

"Riftogenesis regions mark HC deposits.

V.A. Ermakov notes that "the terrestrial crust of magmatic sedimentary origin formed by the middle Proterozoic is a clear evidence of the enormous loss of heat, volatile and easily fusible components of the protomancy. By the end of the period (4.4-1.6 billion years), 85-95% of the continental crust had been formed. The oldest ophiolites are less than 2.8 billion years old. The oldest crustal rocks (protosial - grey gneiss) were formed in the first 500 million years.

O.A. Bogatikov states that "acid rocks contain primary pre-metamorphic zircons, while the rocks of basic composition contain only metamorphic zircons, 1985.

In the Proterozoic (2.5-1.9 billion years) crustal deformation processes take place, accompanied by intracrustal and mantle magmatism and high-temperature metamorphism. By the middle Proterozoic, the crust of magmatic origin has formed (V.A. Ermakov).

The metaporites of the basic and ultrabasic composition are Archean-Proterozoic in age. The first ophiolites are less than 2.7 billion years old.

There is a widespread overlap of the rocks of the greenstone belts with the complexes of the sialic crust.

"In the Archean the axial rotation rate was less than 10 hours" (M.Z. Glukhovsky, V.N. Zharkov, Y.N. Avsyuk), "...due to this in the equatorial latitudes (±35°), under the influence of centrifugal forces in the mantle plume regime, the origin of sialic crust took place [M.Z. Glukhovsky] as well as the formation of the first generation greenstone belts - Barberton and Pilbara (3.4-3.2 billion years)" [Kolger, 2006] [5].

1. The second generation (3-2.7 billion years old) greenstone belts were formed in a rapid axial rotation regime.

Mantle basalt smelting occurred in the Proterozoic (Archean)-Cambrian, Ordovician, Silurian, Devonian, Carboniferous, Permian, Triassic and Paleogene (descending intensity of the magmatic process).

Many researchers [V.V. Belousov, N.L. Bowen, G.S. Gorshkov, B. Gutenberg, N.L. Dobretsov, V.S. Sobolev, V.A. Magnitsky and others] believe that basaltoid magmatism has a "through-crust" character, assuming that the magma generating sources are located within the waveguide. This is confirmed by ultrabasic and eclogitic inclusions ("messengers" of great depths) and the connection of basalt-evidence regions with earthquake sources.

Fluidodynamic processes, (O.M. Borisov), initiate the process of magma formation:

"1) Extra-geosynclinal teleorogenic series - plutons.

Plutonic formations are formed in post-fault time and extend far beyond folding, intersecting the middle massifs and platforms. They are zones of magmatic "revivification", and form deep moving zones. During the Hercynian cycle, 90% of the region's plutonic bodies were formed. Their number decreases from south to north and increases from west to east. According to magnetometry and gravimetry data, it has been established that with depth the adjacent bodies form an ellipsoidal whole with steep contacts. The bottoms of the plutons are located at depths of up to 10 km. The high chemical activity of the magma is indicated by a large number of variously transformed xenoliths and large rims of hybrid rocks.

There is a clear confinement of granitoids to conglomerates, tuffs and tuff-sandstones. The main mass of magmatic bodies, 98.8%, is located in a narrow interval of the geological column (from Ordovician to Upper Carboniferous) within 6-8 km. thicknesses (Tien Shan).

From the laccoliths, apophyses branch off along faults. The roots are confined to major faults. The occurrence of the laccoliths and plutons is controlled by interformational delamination, where they were embedded under high hydrostatic pressure (brachyslip faults are observed).

Calculations of magnetic anomalies [I.A. Fuzailov confirmed the laccolithic morphology of the bodies and their dip to the north.

Most of the bodies have a southward slope.

R.B.Baratov (1973) has established that "Archean deposits of Southwest Pamir and Karategin at first underwent granulite facies metamorphism at T=750o C and P=7 kbar in Karategin and to T=800o C and P=7.5 kbar and higher in Southwest Pamir, further everywhere high temperature diaftorosis and ultrametamorphism under amphibolite facies. Increased pressure has resulted in eclogitization of the rocks. Thus the rocks of crystalline basement have been formed under thermodynamic conditions at T=600-750o and P=6-7 kbar which corresponds to their formation depths of 5-10 km" [5]. [5].

2) The proper-geosynclinal series of plutons is developed only within the movable belt.

3) The extra-geosynclinal volcano-arc series of plutons is developed only within volcanic arcs. In petrochemical features the series is close to volcanics of those tiers, the formation of which it completes. Apparently, the plutons of this series have common focal zones in the Earth's crust and common genetic reasons for close association of volcanics and plutons. Expansion of block borders occurs at the expense of injection of substance masses from deep mantle spheres, and their introduction along the weakened zones of tectonosphere - zones of deep fracture systems, thus volcanic arcs are formed.

4) Injection series plutonic complexes a genetic group that includes ulramafic and mafic, usually subject to serpentinisation.

Bodies of plutons, mafic and ultramafic rocks mark zones of hydrocarbon generation and migration. HC are localized in sedimentary formations of depressions and troughs (HC migration occurs along the weakened apical zone of the pluton, which thickness reaches one km and zones of tectonic disturbances). This process is characterized as hydrothermal" [5].

Primary card Tsvetkov L.D. 2015

Figure A-3. Map of the world basins with estimated shale oil and shale gas formations, as of May 2013 [163].

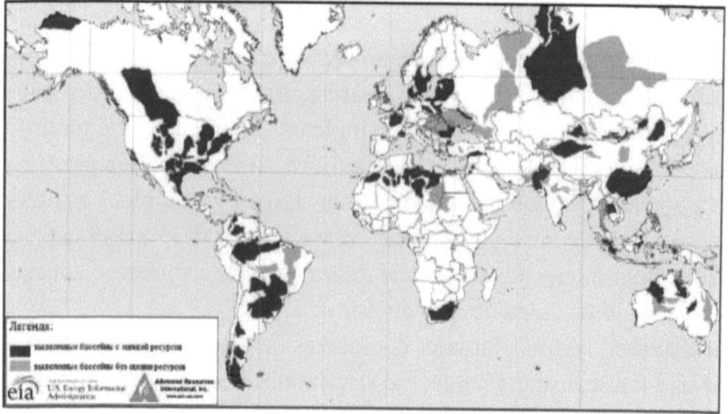

Map of the world's basins with estimated shale oil and shale gas formations, as of May 2013

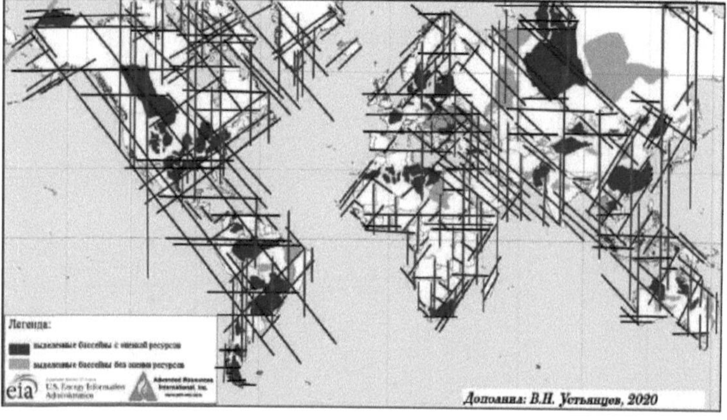

Map. The regular arrangement of structural elements in the Earth system space. Zones of tectonic fault systems control hydrocarbon deposits. Graphical expression of Pierre Curie's principle of symmetry. Compiled by: V. N. Ustyantsev, 2020. A.I. Suvorov established that "north-eastern faults are characterized by thrusts, and north-western faults by shear faults, which articulate at right or obtuse angles and form pairs of faults - "dynamopairs" [5]. The latitudinal and meridional strike-slip faults are characterised as "thrust faults".

"L.I. Ryazanov pointed out the confinement of oil and gas deposits to structural traps of faults active during the latest tectonic stage (Bukhara-

Chardzhou stage, etc.).

M. Valyaev has shown that productive are the intersection nodes of longitudinal and transverse faults, low-amplitude flexure-discharge zones and flexures, branches of intracrustal basement faults whose expression up the section gradually fades, but in all cases the faults are characterized by neotectonic and even recent motions" [8,13].

"Examples of the spatial relationship of terrestrial endogenous deposits with the lines of the global fracture network can serve broad regions on all continents. Here we can mention the location of ore-controlling structures of Eurasia, corresponding to the lines of the global fracture network; the location of ore belts of North America, corresponding to the global fracture network; the location of African tin and diamond deposits, stretching in a straight line chains along the lines of diagonal NE and SE systems of the global fracture network. Hydrocarbon deposits also require supply channels and structural traps for their formation. Both are usually closely associated with a network of linear structures. Spatial connection of oil and gas bearing structures with global fault network lines can be traced in most oil and gas bearing areas of the world. Here one can mention the Barents shelf, Timan-Pechora, Volga-Ural provinces, oil-bearing areas of the Persian Gulf, Bashkortostan, and many others. A number of researchers single out the *submeridional Ural-African riftogenesis belt,* which includes a number of major sedimentary basins with established oil and gas content (A.A. Smyslov et al., 2003) and which belongs to the meridional system of the global network of the 2nd order" [5,7].

Most of the oil and gas bearing structures of the Barents shelf are spatially associated with faults corresponding to the lines of the global fault network. Almost all the largest oil and gas fields and structures, such as Shtokman, Kurentsovskaya, Murmanskaya, Severo-Kildinskaya, Peschanoostrovskaya, etc., clearly gravitate towards the nodes of this network. Another example is the Persian Gulf area, where spatial connection of oil-and-gas bearing fields with SE and submeridional systems of the global rupture network is evident" [5,7].

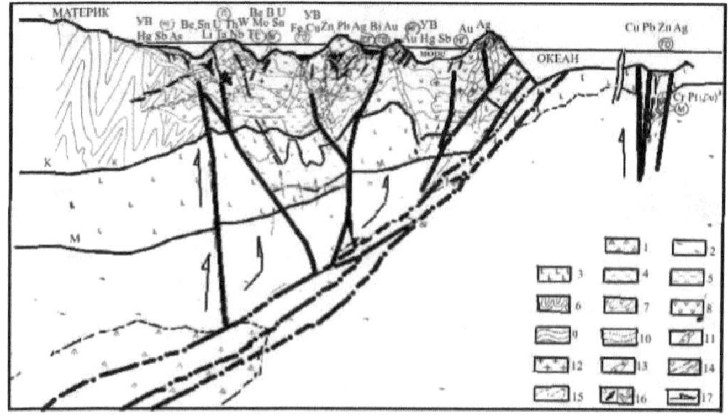

Model of ore deposits formation in the stage of basaltic and granitic magma sources (section) (E.M. Nekrasov).

M - Mohorovitsch surface; K - Conrad surface;

F is the Zavaritsky-Benioff focal zone (subduction zone);

Letters in circles denote deposits: pegmatite - P; skarn - C;

hydrothermal: HH - high-temperature; SG - medium-temperature;

NG - low-temperature; hydrothermal-sedimentary - GO;

I - asthenosphere matter; 2 - upper mantle; 3 - basaltic rocks;

foci: 4 - basaltic magma; 5 - granitic magma; 6 - ancient rocks of granitogneiss bed; 7 - terrigenous-volcanogenic rocks;

8 - extrusive formations;

9 - carbonate-terrigenous rocks;

10 - terrigenous rocks;

II   - intrusives of basic and alkaline composition;

12 - granitoid batholites; 13 - granitoid stems;

14 - deformation zones:

a - subduction (focal);

b - of the deep type;

15 - the inferred boundaries of geological formations;

16 - ore bodies:

a - mined ores;

b - ores of the future;

17 - the direction of fluid flows.

Ore-bearing fault zones. D.V. Rundquist, V.A. Unksov, D.M. Milstein, under ore-bearing fault zones, understand faults both directly hosting ore bodies and controlling their placement. According to morphokinetic characteristics, 7 different ore-bearing associated fault zones are distinguished:

- a system of interconnected linear faults (faults and faults);
- orthogonal systems (splits and transformational shifts);
- steeply dipping faults (mineralisation of predominantly siderophile elements);
- a system of arc-shaped inclined and steeply dipping transverse faults of island arcs and continental margins (Benioff zones with ore zonation (chalcophile element ores at the periphery and siderophile ores towards the centre);
- systems formed by a combination of strike-slip and thrust faults (sometimes turning into pinnacles) and short-period, predominantly fault-slip faults surrounding them, developed along platform margins and foredeep troughs (controlling stratiform deposits);
- systems of associated steeply dipping faults of different orders, manifested as parallel oblique faults and their faults and fault-slip faults (they control many metasomatic deposits of gold, copper, polymetals, five-element formation); swirl" and flexure systems of steeply and declently curving faults as parallel camshaft systems of shifts, strike-slip faults and their faults (vein and greisen, bresite, etc.). deposits of rare metals and gold);
- Radial-concentric fault systems formed by conical (convergent and divergent) and radial faults and spalling fractures (controlling mineralisation of various types, including carbonatites, phenites, feldspatholites, diamond-bearing camberlites)

Supplemented by: V.N. Ustyantsev, 2020.

Западный Кавказ.

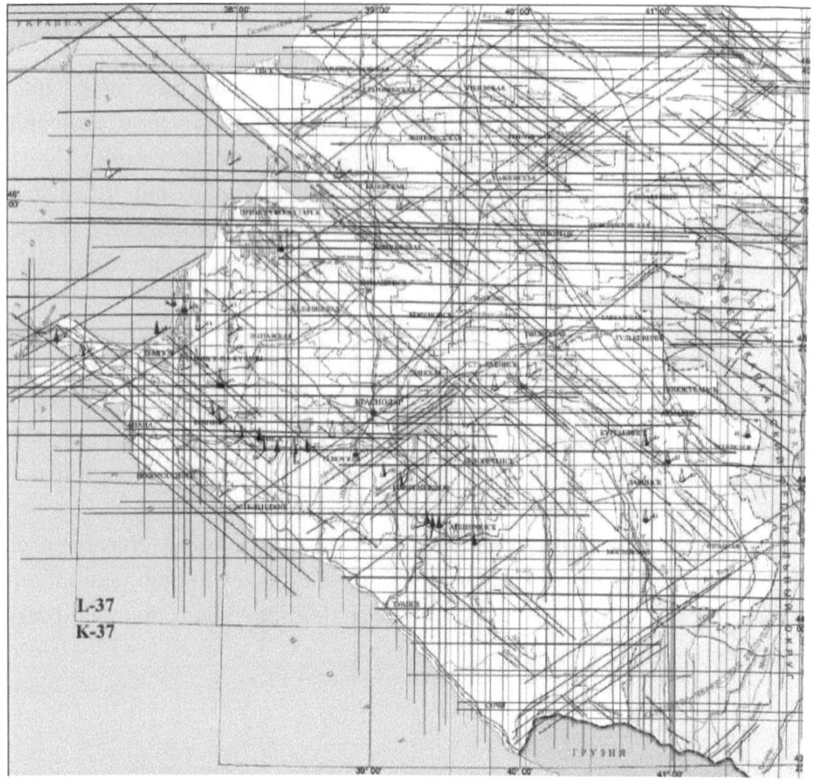

B.B. Tal-Virsky [1972] showed that "heat flux values in Central Asia increase with approach to tectonically active regions and that, geoisotherms often have inverted relief relative to stratopotentials" [8]. This indicates that heat fluxes propagate along guiding structures, which are faults.

M.I. Pogrebitzky, M.V. Rats and S.N. Chernyshev in 1971 showed that "with approaching to rupture the number of cracks increases noticeably, and quite sharply. As the distance from the fracture the intensity graphs

1.   The fractures flatten out and become almost horizontal.

2.   In earlier works, the same authors, based on study of fracturing of rocks of Tajik Depression, Central Kazakhstan and traps Priangarie found that "dependence of distance between neighboring cracks from distance to the break is approximated by exponential function and reminds the picture of stress damping with distance from the earthquake focuses in Reid-Benioff model, and actually observed shifts of fractures like San Aders and others". [8].

3.   ***Paired rifts.***

4.   "Paired faults are commonly referred to as a pair of subparallel linear faults (V.E. Hain, E.E. Milanovsky), between which there is an area of high mobility and permeability, with a peculiar history and complex structure, which reflects the position of the deep fault.

5.   ***Permeable tectonic fault zones***

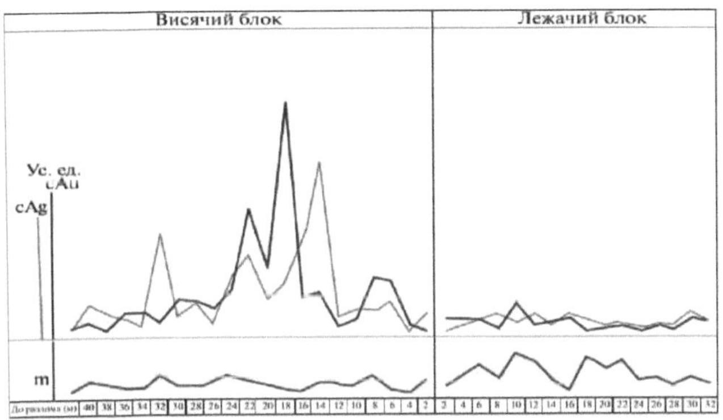

Средняя Азия. Кураминская подзона. Месторождение Кочбулак. О волновых эффектах. Влияние безрудного тектонического нарушения меридионального простирания на распределение концентрации минерального сырья в крутопадающем тектоническом нарушении широтного простирания

Графики построены по данным четырех штольневых горизонтов — 160 м. Концентрация полезного компонента. в рудном теле № 14 определялась посредством про-

On the question of the existence of a wave mechanism:

6.   The mechanism for the emergence of "ore pillars", zones of high permeability of faults; 2. the existence of zones of high permeability extending along faults; 3. the formation of structural "traps". 3. existence of a mechanism regulating the process of degassing of the Earth system. (V.N. Ustyantsev, 1989).

The particular structure of deep faults and their intersection nodes form a closed surface which is an oscillating circuit. The contour is a collector of gases, fluids and magma. Thus, energy-carrying waves entering a heterogeneous medium are reflected and refracted at the media boundaries. These boundaries may cause the appearance of a closed surface from which the waves are reflected which gives the volume, bounded by this surface, vibrational properties and determines the natural periods of the waves, which are typical to the given volume of the vibrational system. In this case the energy of the wave will be given to the transformation of matter. In closed loop conditions the wave velocity will be reduced due to the presence of reflecting surfaces (straight wave propagation is not possible in inhomogeneous medium). Deep fault zone systems are always accompanied by genetically related weakened resonant-tectonic structures, i.e. mineral reservoirs.

The most intense inflow of mantle material is recorded in rifting zones.

"Seismic data record the presence of zones of seismic transparency in the Earth's crust - "zones of absence or significant weakening of reflecting and refracting boundaries", In such zones seismic waves travel with the least loss of energy. Their upper parts do not reach the surface and upper ends can play a role of wave screens, where absorption and transformation (not necessarily thermal) of wave energy will take place" [G.B.Naumov] [5].

A.F. Grachev notices that, "the effect of subslattening is traced to much deeper mantle horizons than Moho boundary, as it has been established for ancient plume of trap province of Parana River. Here a low-velocity mantle anomaly, considered as a result of density deformation associated with the formation of a giant intrusion during solidification of mantle plume material, measuring up to 300 km across, can be traced to depths of 500-600 km". These zones were formed under the influence of the Earth's auto-oscillation system. Deep fault systems, as oscillatory loops, control the migration of matter through the Earth system, the location of energy sources and the shaping of tectonospheric architecture.

Alternation of weakened and dense rocks laterally in lithosphere (R.Z. Tarakanov, N.V. Levy). Fact reflecting lithosphere energy structuring by waves [5,7].

Deforming block stresses using Kamchatka as an example

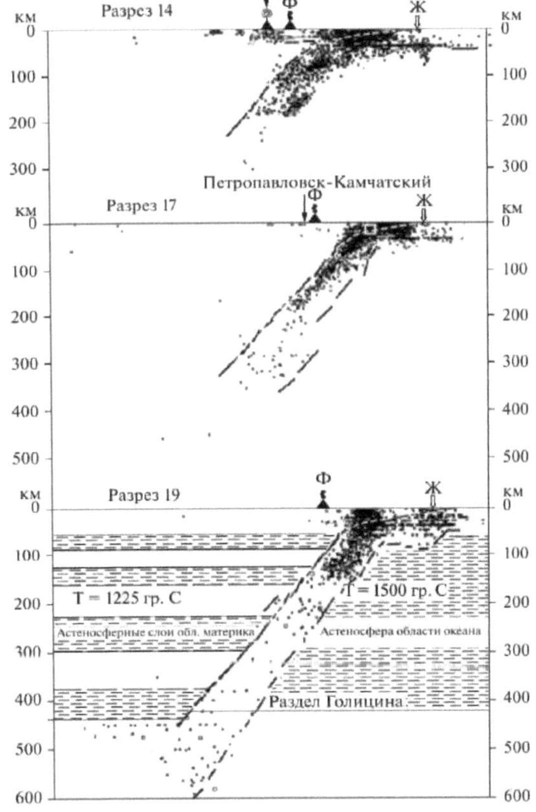

Поперечные разрезы сейсмофокальной зоны
Пунктирной линией показано предполагаемое положение кровли поддвигаемой
Тихоокеанской плиты (по Г.П. Авдейко) *Дополнил: В.Н. Устьянцев, 2020*

"The low-velocity zone in the area of modern volcanism is elevated to a depth of 50 km. Magnetotelluric soundings show that the crust has a layer of increased electrical conductivity at depths of 10-40 km. This layer is developed under the middle part of the peninsula and stretches for 1000 km along Kamchatka, it is linked to an internal volcanic arc. Here the layer is 8-10 km deep and its conductivity is maximal. In the upper mantle a layer of decreased conductivity has been revealed, the top of the layer in West Kamchatka at a depth of 100 km and in the zone of modern volcanism at a depth of 50 km. Towards the

Pacific Ocean the conductivity of the layer decreases considerably (up to some ohms). The surface of the layer is close to the 12000 C isotherm and represents the boundary below which partial melting of matter (asthenosphere) takes place.

It should be noted that conductive zones in the Earth's crust are confined to the geoisothermal range of 400-8000, and rocks at such temperatures have electrical resistivity of hundreds to thousands of Ohm*m.

The nature of conductive zones of Kamchatka with resistance tens-units Ohm* m, is associated with the presence of liquid fluids and electrically conductive sulfide formations" (Y.F. Moroz) [5].

Asthenosphere:

"Thermodynamic calculations of water solubility in silicates on various depths have shown that "retrograde release of water with formation of degassed substance coincides with waveguide" (E.B.Chekaljuk, J.N.Beljevtsev, 1972) [8]. This gives grounds to consider the mentioned layer (asthenosphere) as the main derivative for *volatile and juvenile water extraction. "*During magmatic processes, they migrate mixed with magma and are released during volcanism or *crystallization of intrusives (post-magmatic solutions).* The layer composition is the main source of basaltic magma regeneration (including alkaline magma) and picrites, so it is predominantly basalt-picritic" (V.S. Sobolev, B.G. Lutz) [5,7].

V .A. Magnitsky (1968) [1] in studying the physical nature of the layer found that the low-velocity layer was caused not so much by the effect of high geothermal gradients as by high temperatures and was accompanied by partial amorphization of the primary mantle substance (pyrolite?), but without a significant change in the chemical composition.

"According to I.V. Mushkin's calculations, "the early magmatic stage of alkaline basaltoids (camptonite-terlite-picrete branch of differentiation) of the Southern Tien Shan, took place at 1100-1250°C and 10-15 kbar pressure.

In this range, porphyritic segregations of magnesian olivine, enstatine- and hercynite-rich and chromium-rich protopyroxene, and some spinelides (pleonast, to a lesser extent, chromopycotite) were formed. Decrease of temperature to 1000-1100°C caused inversion of protopyroxene, crystallization of the main mass, formation of magnetite rims around phenocrysts of chromospinelides". (asthenosphere: period of crustal destruction)" [8].

The data of I.A. Efimov 1972, on "eclogites and rocks close to them from the Precambrian of Kazakhstan are of significant interest. He believes that "high pressure (10-12 kbar) and high temperature (6300-6500) are necessary for anthophyllite formation in ultrabasites, which is typical for amphibolite facies conditions. The eclogitic magma is eutectic pyrolite and melted out of the waveguide *at a depth of 50 km. "* "Eclogite-peridotite" sublayer" [8].

The sediment is a derivative of the decomposition of aluminosilicates,

- *the eruptive rocks with which the genesis of oil is associated, i.e., oil,*
- *a mineral of abiogenic origin.*

The likely composition of the layers below (crust, lithosphere:

(From top to bottom)

"1. granulites - 40-50%, migmatites and gneisses - 20-30%, crystalline schists - 10-20%, plagioclasites and granitoids - 10-15%;

2. plagioclasites and gabbro-norites 50-60%, granulites and gneisses 20-30%, granulite eclogites 10-20%;

3. serpentinites 20-40%, eclogitised rocks and eclogites 60-80%;

4. harzburgites and eclogites up to 80%, pyroxenites and lherzolites up to 15%, websterites and gabbro up to 5%;

5. *an amorphised, poorly differentiated basalt-pyrite association [1].*

"High magma productivity, as R. White and D. Mackenzie (1995) note, cannot be provided by melting at the lithosphere level, but requires bringing in material from deeper mantle horizons" [5]. Formation of magmatic formations is accompanied by degassing of rocks and release of (juvenile) postmagmatic solutions which are genetically connected with hydrocarbons. That is, the area of hydrocarbon generation is the lithosphere and the Earth's crust. The localization area is the sedimentary cover of the Earth system. Thus, a complex geochemical system of hydrocarbons (oil, gas).

The processes occurring in the system are associated with the dynamics of geoidal rotation, as indicated by the spatial location of continental roots and the depth of their emplacement, the development of magmagenesis of the equator area, eastern Asia and other regions of the Northern Hemisphere. Degree of differentiation of matter, is reflected by global gravitational negative and magnetic positive anomalies. [5]

S.D. Vinogradov and O.G. Shamina (1968) found a reduced velocity waveguide (Vp =5.7km/s) in the Garm block at depths of 12 to 24 km. Cored waveguides are found at depths of 5.5; 7.0; 10.0; 12.0-24.0 km) Central Asia).

A.N. Dmitrievskiy notes waveguides at depths of 10-25, 55-80, 110-120 km (on the platform - Western Siberia) - fluid-saturated zones have been identified.

T.M. Zlobina notes waveguides at depths of 10-12, 25-28 km, Moho section (Kanimansurkoye province, Central Asia).

The Mohorovičić section has clear boundaries with the higher lying "basalt" layer and is close to the day surface - Aldan Shield, with the lower peridotite layer - the section has no clear boundary.

Far East: 4-8, 11-19, 15-23 km - zones of fluid-magmatic zones (mobile belt). [20]

R.Z. Tarakanov and N.V. Levy - "in the transition zone from the Asian continent to the Pacific Ocean four asthenospheric layers with increased absorption of shear waves, interspersed with layers of increased strength are distinguished in the mantle at depths of 65-90, 120-160, 230-300, 370-430 km.

Central Asia: the asthenosphere roofs beneath the mobile belt are fixed at depths of 80 km, 240 km and 390 km [Lukk and Nersov].

Golitsina section 400-430 km-roof of the upper mantle.

Section: upper mantle soleplate (670) km-middle mantle roof.

Partition: outer core of the Earth-bottom of the lower mantle system - 2900 km.

The concept of neotectonic oil pumping Ragozin L.P. -1980.

"The regularity of HC localization depends on neotectonic phases, which have an oscillatory character. The regular decaying of the intensity of the tectonic process is reflected in the hierarchy of phases; it is explained by the wave nature of tectogenesis and the law of geological retrospective (L.P. Ragozin, 1980). Cycles, phases - reflect epochs of tectonic activity - a high degree of correlation of tectonic activity of WSP with regions of Kazakhstan and N. Asia. Processes of activity and interruptions of ZSP sedimentation - correlated with the inversions of the magnetic field of the Earth system tectonosphere (I.A. Vylshchan, 1969). Maksimov S.P., 1977, showed the relationship between tectonic cycles and the process of oil and gas accumulation - tectonic cyclicity influences hydrocarbon migration. The tectonic environment is a factor controlling the direction and rate of HC migration. At the tectonically active periods, the migration intensifies due to forced action of tectonically active factors such as gravitation, desorption, rock deformation, consolidation of clays, heating of water-saturated strata, increase of oil solubility at heating and increase of capillary forces. Oil and gas migration is considered to be an irreversible process, then the impact of tectonic forces is like a piston pump pushing fluids to their localized locations. As a result of tectogenesis the regular hypsometric and spatial localization of reservoirs, established by B.V. Vityaz and V.V. Bogatsky , 1975, for the WSP. Discrete-periodic localization of oil-and-gas-bearing reservoirs turns out to be a typical property of their hypsometric location, and similar distribution conditions of these reservoirs have interregional significance. In Western Siberia, 4 levels of HC predominant localization are distinguished at absolute markers - 1160, - 11620, - 2060, - 2360, with a step of 400 m between the levels. Lower levels are marked at - 3110 and - 3500 m, higher, at - 800 and - 400 m.

The rejuvenation of the levels is from the bottom upwards.

A retrospective review of the study of tectonic faults
(A. Anokhin)

"In the 19th century, Elie de Beaumont, the founder of the compression hypothesis of the globe, tried to find in the strike of the mountain ranges a certain regular network of directions; he assumed that this network corresponded to the position of the edges of the polyhedron whose shape the globe took when it was compressed.

One of the first attempts to reveal the geometrical regularities of the arrangement of linear landform elements - in this case, mountain ranges - were the works of L. Buch and Elie de Beaumont in the first half of the 19th century. Subsequently, on the basis of the Kant-Laplace cosmogonic hypothesis (on the origin of the Solar System from a cloud of hot matter with its subsequent cooling), Elie de Beaumont put forward the contraction hypothesis (1852). In this hypothesis, folding and mountain building in the Earth's crust - and, accordingly, the pattern of folds and mountain ranges - were explained by crustal contraction due to the cooling and reduction of the Earth's volume.

Further to the end of XIX century the problems of studying structural networks were touched in one way or another by W. Hopkins, D. Phillips, A.P. Karpinsky and others. Many new large linear geological objects, and their sub-parallel aggregates, were identified and tangential forces were cited as the cause of their formation.

At the turn of the 20th century the concepts of formation of the Solar System from cold matter triumphed in cosmogony, which put an end to the contraction hypothesis and many related constructions. At that time E. Süss, studying the location of earthquake epicentres in Lower Austria and Southern Italy, distinguished "seismotectonic lines", on the basis of which W. Hobbs laid the foundation for the modern concepts of systematic lineament and fault networks. In his works 1901 - 1911 this researcher formulated many of the main provisions of the modern concept of regma genesis, in particular, the concepts of lineament and planetary fracturing, the orientation of the main fracture systems in Europe in four main directions. Hobbs' ideas have been developed in the works of I.I. Sederholm (1911), A.P. Karpinsky (1919). In the 1930s, rotational forces (F. Letze, N. Arabyu) began to be involved in explaining the existence of patterns in the distribution of crack systems.

Since the mid-40s, many researchers have come to the conclusion that the same linear structures prevail in different areas of the Earth. The 40s were marked by the works of R. Sonder, J. Umbgrove, G. Stille. The concepts of "regmagenesis" and "regmatic lattice" were formulated, and three main tectonic directions were identified for different continents: sublatitudinal B-direction and two diagonal D-directions ("B-tectonics" and "D-tectonics").

Of the works of the 1950s the works of E.N. Permyakov, N. Butakov, P. Blanchet, J.D. Moody and M. Hill, G.N. Katterfeld. Three types of fracturing were identified:

- local, regional and planetary;

- all new "main" lineament systems (usually diagonal) were highlighted;

- The study of rotational forces as a possible cause of planetary fracturing continued; the structural networks of different planets were compared.

In the *1960s,* with the start of space-based methods of Earth exploration, the amount of work on planetary fracturing increased dramatically.

In 1962 G.N. Katterfeld developed the ideas of the rotational hypothesis and connected not only the origin of the regmatic network but also many other features of the Earth's global relief with the rotational forces. He also formulated the concept of the "critical parallels" at latitudes 35°, 62° and 71°.

Г. Jeffreys in 1963 when comparing the equilibrium figure of the Earth with its observed figure found that the divergence in the polar compressions of these figures provides planetary rotational field stresses of 107 - 108 dyn/cm2, which is sufficient for a very stressed state of the lithosphere.

In 1968, A.I. Suvorov came to the conclusion about the predominant inheritance of tectonic structures, about planetary meridional compression as the main cause of faulting along the four main directions (submeridional, sublatitudinal, NE and NW).

Massive measurements of directionality of relief lineaments on small-scale maps were made in the 60s by P.S. Voronov (1968) in co-authorship with S.S. Nezametdinova. The resulting rose diagram of linear structures for the entire Earth's landmass reflects the symmetry of the network of these structures relative to the planet's axis of rotation, as well as the presence of four diagonal systems.

In the 1960s and 1970s, a number of scientists continued studies of planetary fracturing; the works of P.S. Voronov, I.I. Chebanenko, K.F. Tyapkin, S.S. Shultz (senior), A.N. Lastochkin should be noted from the works of this period.

In 1975-76, U. Y. Carrey justifies the idea that the Earth's radius has increased by almost 2,000 km in the last 2.75 billion years, from 4,400 km to today's 6,378 km.

The rotational hypothesis was further developed in the works of M.V. Stovas (1975), where the author justified many tectonic phenomena on the Earth and other planets by changing their shape due to long-term slowing down of rotation caused by tidal forces. *E. Kanasevich et al (1978) showed a high degree of spatial organisation of the planetary face for the entire Phanerozoic.*

In the 70s and 80s, a number of comparisons were made *between* the planned location *of lineaments and the distribution of mineral deposits.* Here the works of I.N. Thomson, M.A. Favorskaya, I.K. Volchanskaya, S.S. Schultz Jr.

In *1983, a book* entitled 'Space Information in Geology' *was published*, in which a team of authors presented numerous structural constructions in various regions of the world based on *space-photography* data.

Particularly noteworthy is the work of A.V. Dolitsky (1985), who states the existence in the Earth's crust of a planetary network of ruptures of 4 fixed directions - N, NE, E, NW; different-aged variants of this network are distinguished, oriented relative to different positions of the moving poles.

In 1985-86, works by J.G. Katz, A.I. Poletaev and E.F. Rumyantsev, in addition to describing regional lineament networks, suggested that lineaments are natural indicators of the earth's crust division lines.

In the 1980s, many researchers came to the conclusion that the Earth is substantially symmetric. A number of features of the Earth's structure were described that speak of its symmetry, such as the *uniform distribution of mid-ocean ridges, island arcs and other large forms* of relief of submeridional strike-slope, *about 90° apart.* In this connection it is necessary to note the works of Ch. Pan (1985), V.N. Sholpo (1986), E.E. Milanovsky, A.L. Nikishin, (1988), G.F. Ufimtsev (1988).

In the 1990s, I.I. Chebanenko, M.L. Kopp, V.S. Rozhdestvensky, P.S. Voronov, L.M. Raszvetsev and many others *published the results of their research on this subject.* In addition to regional constructions, their works also contain ideas of the global level (e.g., introduced by P.S. Voronov *the concept of geofluction - the tendency of "flowing" of crustal matter towards the equator under the action of centrifugal forces.*

In *1991 - 1993 the three-volume book* "Faulting in the lithosphere" edited by N.A. Logachev *is published.* The authors of the three-volume book, including S.I. Sherman, considered in detail physical, structural and other aspects of faulting under tension, compression and shear stress.

In 1994 A.I. Poletaev substantiates the significance of the notion of "lineamentary division of the Earth crust", proves the existence of different-scale networks of lineaments throughout the planet, connects with them the development of geological processes, distribution of mineral deposits. E.E. Milanovsky in 1995 gave a new development of the idea of the Earth expansion (pulsation).

In 1996, *a new Geological Atlas of Russia at a scale of 1:10,000,000 was published*, edited by A.A. Smyslov, which includes a map of cosmogeological objects of Russia, *entirely devoted to lineaments,* among a large number of maps of geological content.

Rich material for comparing the structural plans of the Earth and other planets is contained in the 2000 edition of G.N. Katterfeld, which contains numerous photographic illustrations of various regions of the Earth and planets. Quantitative information about the directionality of linear structures is summarized in quite a few rose diagrams, which greatly facilitates their comparison.

In 2004 IWGSEE published the volume "Tectonics and Geodynamics" of the encyclopaedic reference book "Planet Earth", edited by L.I. Krasny. This volume contains a number of articles that cover the current level of scientific knowledge about the structure of the Earth, including global tectonic concepts that, to a greater or lesser extent, relate to planetary structural networks. Among the authors of the volume who published articles on the topics close to the topic of this work, we should mention L.I. Krasny, B.A. Bluman, E.E. Milanovsky, G.F. Ufimtsev, Y.M. Pushcharovsky, N.I. Pavlenkova, Y.N. Avsyuk, S.I. Andreev, A.H. Kagarmanov, E.M. Pinsky, A.K. Khudolei" [5].

По М.В. Муратову, 1975:

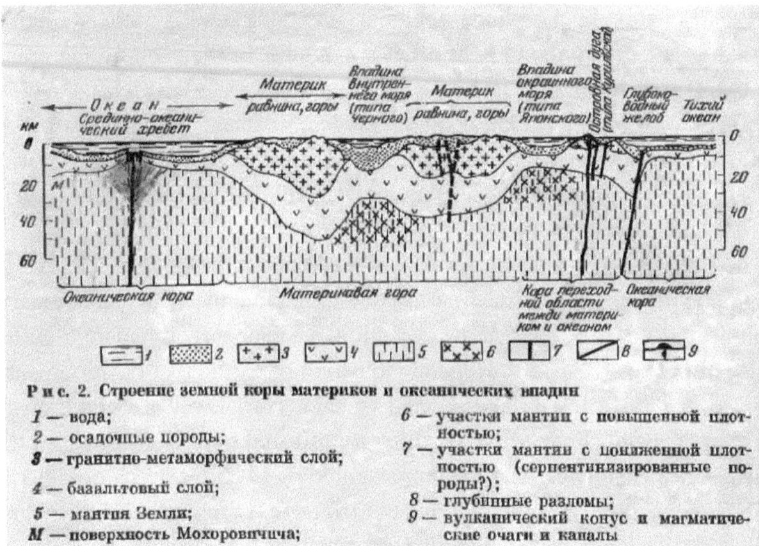

Р и с. 2. Строение земной коры материков и океанических впадин

1 — вода;
2 — осадочные породы;
3 — гранитно-метаморфический слой;
4 — базальтовый слой;
5 — мантия Земли;
М — поверхность Мохоровичича;
6 — участки мантии с повышенной плотностью;
7 — участки мантии с пониженной плотностью (серпентинизированные породы?);
8 — глубинные разломы;
9 — вулканический конус и магматические очаги и каналы

On the multidirectional process of substance migration

According to V.I. Schreibmann's calculations:

- "isostatic compensation is achieved only at levels close to the base of the upper mantle (the area of the roots of mountains and continents)" [Belyaevskiy, 1974] [1]. This testifies to its density heterogeneity, which as a whole is decompacted under orogens and only within the Fergana depression its compaction is observed. Here excessive density of the upper mantle substance reaches 0.5 g/cm3. [Butovskaya, 1977, Fergana depression] [5,7].

Seismic surveys show the presence of bulges beneath the depressions, where the properties of matter change.

The formation of bulges under depressions is most naturally linked to the mobility of seismic segments and to the acquisition of mantle properties by rocks that were previously part of the crust.

This fact indicates the presence of downward flows of sedimentary material, which, as it sinks, takes on the properties of mantle rock.

"The teleseismic tomography study assumed that the lateral heterogeneity is concentrated in the layer from the Earth's surface to a depth of 300 km. The strongest velocity inhomogeneities were found to be immediately below the Earth's crust.

The anomaly was detected *by researchers at the Institute of Physics of the Earth, USSR Academy of Sciences,* by calculating isostatic anomalies of gravity averaged over $^{10 \times 10}$ areas, and is caused by extensive density inhomogeneities at great depths.

Against this background, regional anomalies with rather significant horizontal gradients up to 0.15 mLg/km are exhibited, their amplitude reaching several tens of milligal. The largest negative anomalies cover Central Asia with B=-1 density, with layer thickness (anomalies) more than 500 km. in Pamir-Alay, 350-500 km. in Northern and Southern Tien Shan, Bukhara-Gazli and Mari, and 150-300 km. in Fergana valley and Turan plate. (IFZ, RUSSIAN ACADEMY OF SCIENCES).

In 1979, S.I. Ibadullaev and K.K. Karabayev in their paper "On evolution of magmatic process in Central Asia", based on factual material (geological map of Central Asia (1976)), showed the evolutionary stage of magmatism in different periods (from Proterozoic to Neogene inclusive) of earth crust development and concluded that "all known intrusive and volcanic complexes in Central Asia are derivatives of magmatic processes, **which manifested twenty eight times (from Proterozoic to Neogene).** They are represented by seventeen rock complexes of different composition, genesis and time of formation. Differentiation of magmatic formations took place in the direction of: alkaline - acidic - basic - ultrabasic rocks.

The frequency of occurrence of magmatic complexes varies from 1 to 16. Thus, leucocratic, biotite and dicloudite granites, granodiorites, granite-gneisses were intruded 16 times (Archean-Neogene); gabbro, norites, gabbro-diorites, diorites 14 times; rocks of the granodiorite, quartz diorites, granite-gneisses and granite-diorite-gneisses complex 13 times; diorites, gabbro-diorites, quartz diorites, quartz syenite-diorites - 11 times; dunites, peredotites, serpentinized harzburgites - 5 times (in the Cambrian, Ordovician, Devonian and Carboniferous); complex of rocks - peridotites, pyroxenites, gabbro, gabbro-norites - 1 time (Cretaceous). The complex of gabbro, gabbro-norite, which corresponds to "basalts" was intruded 14 times (from Archean to Neogene inclusive).

The complexes of acidic and basic rocks have a high frequency of intrusion, while the series of alkaline and ultrabasic rocks have a lower frequency.

In each individual period, differentiation took place towards a change in the composition of the magma from acidic to basic" [5].

Sediments are derived from the decomposition of aluminosilicates, the igneous rocks that are associated with the genesis of oil, i.e. oil is a mineral of abiogenous origin.

The likely composition of the layers below (crust, lithosphere:
"1. granulites - 40-50%, migmatites and gneisses - 20-30%, crystalline

schists - 10-20%, plagioclasites and granitoids - 10-15%;

2. plagioclasites and gabbro-norites 50-60%, granulites and gneisses 20-30%, granulite eclogites 10-20%;

3. serpentinites 20-40%, eclogitised rocks and eclogites 60-80%;

4. harzburgites and eclogites up to 80%, pyroxenites and lherzolites up to 15%, websterites and gabbro up to 5%;

5. an amorphised poorly differentiated basalt-pyrite association [1].

"High magma productivity, as R. White and D. Mackenzie (1995) note, cannot be provided by melting at lithosphere level, but requires bringing of material from deeper mantle horizons" [5]. Formation of magmatic formations is accompanied by degassing of rocks and release of (juvenile) postmagmotic solutions with which hydrocarbons are genetically connected.

"The latest Neogene-Quaternary post-platform horogenic stage. In southern Tien Shan, high seismic activity is manifested, and in the north, vaulted rift uplifts and their dissecting faults and grabens are formed and attributed to the Trans-Asian Nalivkin Belt. The epoch was accompanied by the rise of heated water with a number of metals and volatile compounds of mercury and antimony dissolved in it. Heated oil and near-surface waters also circulated. They additionally redeposited and concentrated *gas, oil, sulphur, strontium, ores of non-ferrous metals, number of rare and disseminated elements* in sedimentary formations of young Mesozoic and Cenozoic covers" *[5].*

Oil is of polygenic origin.

Under the influence of the force of gravity directed towards the centre of the Earth system, the planet took on the shape of a sphere. A global stress field was created, which was unloaded by a global network of tectonic faults both radially and laterally, from the Earth's surface to the centre of the system, facilitated by centrifugal forces of rotation. The process of displacement of primary abiogenous fusible volatile elements and their compounds from deep spheres of the Earth system into the crust of magmatic origin is connected with these acting factors (gravity and centrifugal force of rotation).

"Undoubtedly, under conditions of high pressure and magmatic masses, the formation of many petroleum hydrocarbons is possible.

It must be stressed that oils cannot be regarded only as hydrocarbons. Hydrocarbons only predominate in their composition. They always contain many percent, sometimes tens of percent, of compounds containing O, N, S. The explanation of their genesis cannot be based only on the explanation of the genesis of hydrocarbons. *It is often forgotten"* (V.I. Vernadsky, 1934)

Primary hydrocarbons and oil of abiogenous origin:

"Primary - independent of the biosphere - carbon minerals" (Vernadsky,

1934):

These are the accession products of calcium cancritites to aluminosilicates: 3Na2 Al2 Si2 O8 * C^C^X

- ***Eruption rocks rich in cancrinite,*** containing up to 1.7% CO2, (0.74% C). Eruption rocks with juvenile calcite are even richer in carbon - trachyte from Bilbao contains 2.09% C, phenite from Norway (by Bregger) contains 9.6% C.

- ***Oxygen-poor rocks*** include, CO, SCO, NSNO (formic aldehyde), HCOO (formic acid). These bodies are formed at high temperatures by the reduction of carbonic acid in the presence of water and hydrogen sulphide. They are not very rare, but occur only in the trace state in juvenile and phreatic gases.

- ***Oxygen-deprived minerals.***

- ***Among the carbonaceous minerals devoid of oxygen,*** the most important are hydrocarbons - CH4, C2H6, etc., metallic carbides and native carbons.

The genesis of carbon monoxide is, for the most part, independent of carbonic acid. The presence of iron and nickel carbonyl compounds in the Earth's crust indicate this.

"Gases found in volcanic eruptions: water, hydrogen, methane, carbon monoxide and carbon dioxide, formic acid and CSO" [1].

"Free hydrogen is found in large quantities in magmas and erupted rocks. When water and carbonic acid act in the depths of the Earth's crust, significant masses of CO can be formed. CO, in the atmosphere is in negligible amounts - it does not accumulate one way or another" [1].

"Free nitrogen, corresponding to carbonic acid in the geochemical history of carbon, is the main juvenile mineral for this element. It is stable in all known shells of the Earth's crust" [V.I. Vernadsky, 1934].

V.I. Vernadsky:

"Under the thermodynamic and chemical conditions of the Earth's crust, all natural carbonaceous compounds are stable if protected from the influence of life.

Free nitrogen, corresponding to carbonic acid in the geochemical history of carbon, is the main juvenile mineral for this element. It is stable in all known shells of the Earth's crust.

Free hydrogen is found in magmas and in erupted rocks in large quantities.

When water and carbonic acid act, significant masses of CO can form in the depths of the Earth's crust. CO, is negligible in the atmosphere - it does not accumulate one way or another.

We see that CO, CO2, CH4, can be formed under terrestrial conditions at the expense of other juvenile bodies whose presence is certain (C, CO2, H2O, iron compounds).

Undoubtedly, the formation of many petroleum hydrocarbons is possible under conditions of high pressure and magmatic masses.

It must be stressed that oils cannot be regarded only as hydrocarbons. Hydrocarbons only predominate in their composition. They always contain many percent, sometimes tens of percent, of compounds comprising O, N, S. The explanation of their genesis cannot rely solely on the explanation of the genesis of hydrocarbons. This is often forgotten.

Petroleum, must be considered minerals of biogeochemical origin, which have undergone strong metamorphism" (V.I. Vernadsky, 1934). V.I. Vernadsky proved that: "carbon compounds participating in the structure of caustobioliths, including petroleum, represent an integral part of the geochemical system of the carbon cycle in the Earth's crust, in which the main role belongs to the living matter of the biosphere".

B. I. Vernadsky proved: "the ability of living matter, including lower organisms, to concentrate enormous reserves of organic carbon in the lithosphere".

- He was also the first to determine that: "hydrocarbon gases, unlike oil, arise from OM of all types".

One of Vernadsky's important definitions is that

"the oil accumulated in the fields is a very small fraction of the total mass of oil in non-reservoir formations". [5].

"Only a fraction of the matter of organisms collects as caustobioliths - 0.3%. It is only that part which escapes from the circle of life. This insignificant part, however, appears in nature around us in huge masses of oil, or hard coal, in hundreds of millions of tons of carbon" [1]. V.I. Vernadsky (1934), writes:

"Cyclic elements make up almost all the mass of the Earth's crust - 99.7;%. The remaining small residue, 0.3%, is not insignificant. It amounts to quadrillion metric tons. There are radioactive elements concentrated in it, which are of great importance in the life of the biosphere. This is matter in chemically active state, which has free (atomic) energy, producing in the Earth's crust enormous chemical work. The amount of such matter is 10^{15} tons. Close to the same order of magnitude is the mass of another "active" matter - living matter (living organisms), no less deeply imbedded in the mechanism of geochemical processes. There are two types of "chemically" active matter in the Earth's crust: radioactive elements and living matter - a collection of living organisms". "The elements are found in siliceous-aluminous masses - complex, ever-changing systems, more or less viscous, with high temperature and high pressure, overflowing with H2O-CH4 gases. The genesis of these atmospheres must be complex, water vapours and hydrocarbons may be of different origin" [V.I. Vernadsky, 1934]. "The formation of petroleum is one of the manifestations of the great importance of the process, the transfer of energy from the sun through living matter to the deep layers of the planet" [V.I. Vernadsky, 1934].

Нефть (кероген) фанерозоя концентириуется обогащая кровлю или подошву НГМТ

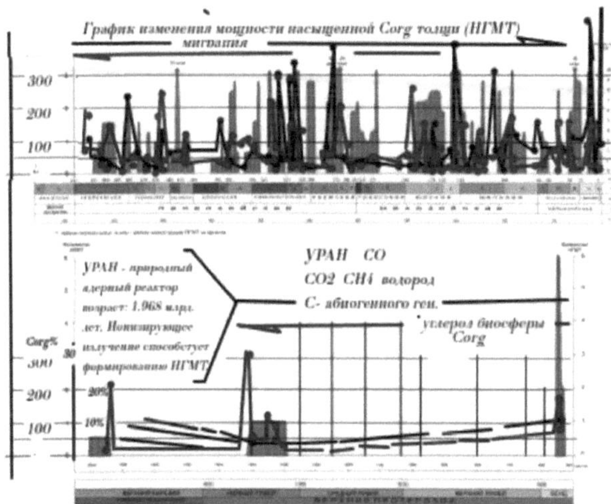

Составил: В.Н. Устьянцев, 2020

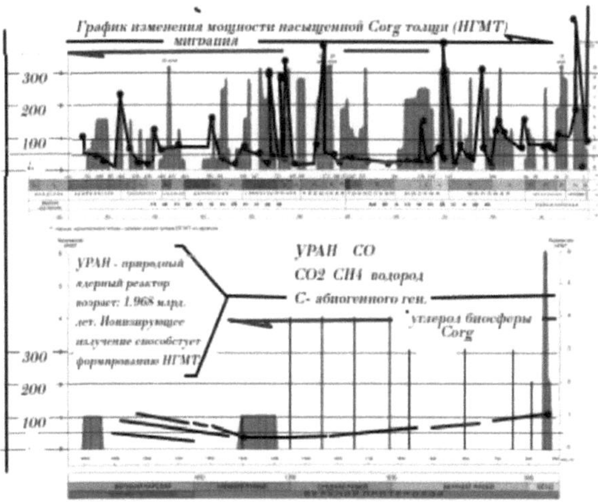

Составил: В.Н. Устьянцев, 2020

Факторный анализ стратиграфического разреза мира (НГМТ) Прогнозны построения на основе фактматериала. (В.Н. Устьянцев, 2020):

With metamorphism, the proportion of C increases and the proportion of H and hetero-elements decreases.

The change in the thickness of the Corg-saturated NGMT layer is a vivid manifestation of the effect of the structural-material transformation of the Earth system under the influence of energy waves. "Peaks", - fix the zones of stretching of the Earth's crust. Deep fault zone systems are always accompanied by genetically connected with them weakened resonance-tectonic structures - reservoirs of minerals. In this case, the Corg enriches the roof or the basement of the NGMT. In the Phanerozoic, cycles of crustal destruction are clearly distinguished.

"Goldwyer Formation (O21). The highest organic matter enrichment (6.4%) is characteristic of the upper part of the section of the Goldwyer Shale" [25].

The Khabour Formation (O2-3) is described in northern Iraq and has a terrigenous composition. According to Akkas-1 drilling, the lower 500 m of the more than 2100 m thick section is black shale with a moderate total organic carbon (TOC) content of 0.9 to 5%.

Organic matter initially migrates into the Earth's crust in a diffuse form and only thanks to structuring energy waves of both longitudinal and transverse type, generated by the Earth's auto-oscillation system, conditions favorable for processes of concentration, differentiation of useful component, with manifestation of geochemical zoning, are created in the tectonosphere. Otherwise, the substance would remain in a dispersed state.

The formation of hydrocarbon deposits of both simple and complex composition occurs continuously from the Archean to the Quaternary. This process reflects one of the most important systemic properties of the planet, which is expressed by the preservation of its dynamic equilibrium in the space of space. The Earth's self-oscillating system, in its cyclically directed evolutionary development, is not destroyed, but, passing from one energy state to another with each cycle, undergoes a process of transformation at the atomic level.

Sediments are derived from the decomposition of aluminosilicates, the igneous rocks that are associated with the genesis of oil, i.e. oil is a mineral of abiogenous origin.

"Absolute maximum intensities *were found to* fall in the following intervals (billion years): 2.65-2.60; 1.90-1.85; 1.00-0.95; 0.55-0.50 and 0.30-0.25. *If we exclude* the interval 0.55-0.05, the others are 0.8+_0.1 billion years apart, i.e. they form a quasi-regular cyclicity. On the other hand, the 0.55-0.50 peak dropped out of this sequence, together with weaker peaks of the second order, form another series: 1.2-1.15; 2.1-2.05 and 2.85-2.8. They coincide with the final phases of the most intense growth pulse of the juvenile continental crust in Earth history. *The process occurred in a wave-like manner" [5].*

Large provinces of igneous rock formations (LIPs), potential sources of oil and gas deposits.

The leading factor in the formation of mineralogical associations is the energy (cosmogenic) factor, which determines the mechanism and conditions of formation and genesis of minerals.

"The simultaneous manifestation (according to V.V. Belousov, 1975), on the surface of continents of different endogenous regimes, "indicates the heterogeneity of the Earth thermal field: at the same time heat flows in different places differ in their intensity, hence heat flows change their intensity both in space and in time" [4].

"This fact indicates the existence of a single control mechanism, under the influence of which the system and objects, in its geological space, evolve. The Earth's tectonosphere system is a complex energy system, the state of which is determined by geological processes and the physical-chemical deformations that occur between the constituent elements of the system, (the acting factor of tectonosphere formation is cosmogenic).

There is a temporary lag of the hydrothermal ore-forming process and localisation of minerals of all types, in fracture-breccia, all morphological types of structures - flat, steep-dipping, tubular, flexural, etc., in which the following occurs: gold, uranium, strontium, mercury, oil, gas, gas condensate and others (deposits of Central Asia, Western Siberia).

This circumstance is explained by the difference in mass-flow migration velocity - fluid, and the velocity of the energy wave, under the influence of which the cyclic-directional structuring of the geological space of the Earth system takes place.

Time gap between structuring of tectonosphere by energy wave and hydrothermal mass flow, formation of granitoid massifs, is about 50 mln years. This process is characterized as directional cyclic (wave-like).

The correlation of the processes of ore formation with the manifestation of the epoch of peneplanation reflects the presence of a single wave mechanism of structure formation and ore formation; the unity of the global and regional, as well as the cyclic nature of their manifestation in the history of the Earth system. The process of matter migration occurs both in the direction of the core and vice versa, i.e. it has a multidirectional character. This is a fundamental position in understanding the process of ore formation and the genesis of mineralogical associations.

By migrating from one formation to another, matter undergoes transformation at the atomic level, acquiring new qualities and properties. Physical-chemical deformations are genetically connected with interacting stress fields, the occurrence of which is related to the energy of the feeding systems of higher organisation.

Along deep faults, there are genetically related resonance tectonic structures - reservoirs of minerals.

The formation of oil deposits occurs over a period of 50 million years and correlates with intensifications (epochs) of magma genesis.

"Petroleum, must be considered minerals of biogeochemical, polygenic origin, which have undergone strong metamorphism" (V.N. Ustyantsev, 2020).

Due to the fact that:

Elemental composition of oil: C 82.5-87%; H 11.5-14.5%; O 0.05-0.35, rarely up to 0.7%; S 0.001-5.5%, rarely over 8%; N 0.02-1.8%. About 1/3 of all oil produced in the world contains more than 1% S.

Mean Corg in the stratigraphic section of the world: Corg=5%, n=50 formations from Paleoproterozoic to Quaternary analysed. (V.N. Ustyantsev, 2020).

The hydrocarbons are complementary to each other - methane+oil. The generation potential of hydrogen+methane is juvenile and has high values:

- hydrogen *index H1 = 250-500 mg*, at ^^=9%, (Lower Riphean);

- Neogene: - H!-750, at Corg=2%) obviously determines the potential for hydrocarbon deposits.

Change in Corg max: Proterozoic 29%; Quaternary: 1%. Average Corg across the stratigraphic section, - 5%. Mean: HI = 361.5. Mean:(81+82) = 1.39. K = ((S1+S2)/HI)*100 = 0.4%.

I.e., the value (99.6%) indicates that huge masses of minerals were formed by the HI, the deep crustal spheres of the Earth's system.

"Free hydrogen is found in large quantities in magmas and erupted rocks. When water and carbonic acid act in the depths of the Earth's crust, significant masses of CO can be formed. CO, in the atmosphere is in negligible amounts - it does not accumulate one way or another" [1].

"Free nitrogen, corresponding to carbonic acid in the geochemical history of carbon, is the main juvenile mineral for this element. It is stable in all known shells of the Earth's crust" [V.I. Vernadsky, 1934].

The formation of fields with large hydrocarbon reserves is due to carbon not related in its origin to the biosphere (juvenile) and the high generative hydrogen potential of HI.

The content of kerogen in oil shale is up to 60%, mostly 1535%.

Elemental composition of kerogen in the catagenesis zone (%):

- sapropel type - C 64-93; H 6-10; O 0-25; N 0,1-4,0; S 0,1-8,0;

Considering that the average Corg=5%, there is 59-88% abiogenous carbon in the kerogen.

- humus-sapropel type - C 64-96; H 1-5; O 3-25; N 0.1-2.0; S 0.13.0.

Considering that the average Corg=5%, there is 59-91% abiogenous carbon in the kerogen.

With metamorphism, the proportion of C increases and the proportion of H and hetero-elements decreases.

The authors (Gottich R.P., Pisotsky B.I. et al., 2012), conclude

"The oils of all oil and gas bearing provinces are enriched in relation to the upper crust clark in elements characteristic of volcanic fumarolic gases: Hg, As, Sb, Se, Te, Cd, Ag, Au; selectively: Re, Ni, Cr, Pb, Bi;

- platinum group elements are present in all oil samples, distinguishing the oils from the upper crustal complexes and sedimentary cover rocks;

- in platinum-bearing ores isolated with isotopic dilution from oil, the spectrum normalised to chondrite is similar to that of platinum-bearing hyperbasite ores;

- provincial oils differ both in the set of elemental series in the diagrams: even-odd, and in some indicator ratios: Ru/Ir, Ti/Y, Zr/Nb, Nb/Ta, Th/Yb;

- chondritic-normalized spectra of oil lanthanides, for the most part, are characterized by a pronounced positive anomaly on europium, which distinguishes them from similar spectra of organic matter, formation (buried) waters, sedimentary and acidic magmatic and metamorphic basement rocks. The slope angle in normalized spectra of lanthanide oil (La/ Yb)N is apparently determined by both alkalinity of the source and the value of fluid pressure" [5,7].

Kerogen: - The structure of a kerogen is represented as a macromolecule composed of condensed carbocyclic nuclei connected by heteroatomic bonds or aliphatic chains.

Petroleum is a complex natural formation consisting of hydrocarbons (methane, naphthenes and aromatics) and non-hydrocarbon components (mainly oxygen, sulphur and nitrogen compounds).

Example:

"The Pingdiquan Formation (Lower Wuerhe, Lucaogou to the south) (P2) is common in the Junggar Basin of western China. The Sorg content is 5% on average, up to a maximum of 20%. The rocks are formed in lacustrine-lagoonal and fluvial-delta environments. Kerogen type I and II, no coals. It is one of the richest oil-and-gas-bearing formations in the world, enriched with fluid fluids. The sediments are up to 1,200 m thick and consist of grey and black mudstones and dolomitic mudstones interbedded with thin layers of sandy mudstones, schistose clayey siltstones, siltstones and fine sandstones" [2,7].

Average Corg in the stratigraphic section of the world: Corg=5%, array of formations - Paleoproterozoic - Quaternary: n=50. Oil contains 77.5-82% carbon of abiogenous origin (not genetically linked to the biosphere).

Thus, the polygenic origin of oil must be considered proven, qualitatively and quantitatively.

All hydrocarbons are complementary to each other.

Petroleum, should be considered minerals of biochemical, polygenic origin that have undergone strong metamorphism. (V.N. Ustyantsev, 2020).

The origin of kerogen (oil) is genetically connected with the work of the Earth's auto-oscillation system, i.e. with its energy, which increases with time (acting factor: the process of self-organization of the auto-oscillation system). The entropy of the system decreases with the course of geological time, and the amount of Corg, - increases, due to processes of metamorphism of oil (kerogen). At the same time, the amount of hydrogen H, - decreases.

Petroleum, in the process of self-organisation of the Earth system over geological time, undergoes significant metamorphic transformation - the free hydrocarbons are degraded. The amount of carbon increases and the amount of hydrogen decreases.

This process is largely due to the decomposition of silica-aluminosilicates, which, as shown by Vernadsky (1934), contain large quantities of abiogenous carbon.

"The average carbon content in the Earth's crust, according to A.E. Fersman, is 0.35% (1939). It is determined on the basis of numerous analyses of rocks, natural waters and air. It is clear that such estimates are far from accurate, and the data from different authors differ significantly. However, the order of values of both the total carbon content and its distribution in different zones of the Earth's crust seems to be correct" (V.I. Vernadsky).

"The importance of an element in the life of the Earth is by no means determined by its abundance. For example, carbon in the Earth's crust is only 0.1% (by weight). Despite such a negligible content, it plays an extremely important role in the life of the Earth. All biological processes are directly related to carbon. This element is the basis of life on Earth.

Carbon has the ability to attach atoms of different elements - it forms up to three million compounds of all kinds" (Chemist's handbook).

The systemic properties of carbon, contribute to the formation of mineralogical associations in the structured energy wave tectonosphere of the Earth's auto-oscillatory system.

Conclusion:

Average Corg value in the stratigraphic section of the world: Corg=5%, n=50 formations from Paleoproterozoic to Quaternary were analysed. The Corg=5% parameter is key; it objectively, regardless of our perceptions, defines the oil genesis as polygenic.

Abiogenic and biogenic hydrocarbons are complementary.

The author's analysis, without touching on hypotheses or theories, made it possible to quantify the genesis of oil, which will enable scientists to focus their efforts on developing technologies for hydrocarbon deposits with a low oil recovery factor.

"Combined hybrid processes may also be of interest because they can simulate and implement self-cleaning processes occurring in natural environments. Under natural conditions, the transformation and decomposition of organic substances of biogenic origin and anthropogenic organic pollutants, as well as the natural self-purification of ecosystems occur, as a rule, as a result of simultaneous biological, chemical and photochemical processes In abiogenic transformation of compounds, in particular, the important role of hydrogen peroxide, metals of variable valence (Re, Mn), titanium containing minerals, ultraviolet component of solar radiation is known. The leading role in the biogenic transformation of various organic compounds in soil and aquatic environments belongs to microorganisms" (p.230) [6].

The wave mechanism of mineral concentration in blocks of the Earth's crust:

1. the Earth's self-oscillatory system and the genetically related hierarchy of second-order self-oscillatory systems (structural elements), determine the existence of a single mechanism under the influence of which all types of minerals are concentrated (factor - favourable PT conditions).

2. Minerals (of any type) are confined to intensely dislocated strata, i.e., compression zones, and, within them, to local tensile zones (fracture-breccia structures). In this case, the repeated change of compression conditions by tensile conditions is capable of leading to a high concentration of minerals.

The mechanism operates under the influence of the Earth's auto-oscillation system.

Regularly spaced zones of intense permeability are structural barriers to all types of minerals (PT factor) and reflect epochs of crustal deformation by endogenous processes and wave energy.

The mechanism of structuring by the wave energy of the Earth's tectonosphere's auto-oscillatory system, leads to its volumetric expansion and the appearance of periodically discretely located weakened zones, which are performed by volatile and highly charged elements. Such zones concentrate scattered hydrocarbons, which, due to their geochemical features, cannot stay in the weathering crust for a long time - they are only nucleated in it. The hierarchy of energy waves initiates the process of migration, differentiation of mineralogical associations, i.e. promotes their concentration in weakened zones of different morphology both laterally and radially. The process of formation of high permeability energy zones, both laterally and radially, is genetically connected with the energy wave generated by the Earth's auto-oscillation system.

The density heterogeneity of the energy divisions of the Earth system tectonosphere reflects the cyclic nature of its structural-matter transformation.

Organic matter initially migrates into the Earth's crust in a diffuse form and only thanks to structuring energy waves of both longitudinal and transverse type, generated by the Earth's auto-oscillation system, conditions favorable for processes of concentration, differentiation of useful component, with manifestation of geochemical zoning, are created in the tectonosphere. Otherwise, the substance would remain in a dispersed state.

"Any open, dissipative and non-linear systems inevitably experience self-oscillatory processes, supported by external sources of energy, resulting in self-organisation" [I.R. Prigogine].

"The efficiency of the influence of structural factors, as well as external - the Moon and the Sun - constant energy sources, on the hydromagnetic dynamics of the core is very small and difficult to estimate" [S.V. Starchenko, 2009]. This fact suggests that the Earth system is auto-oscillatory, due to the fact that it operates at the expense of its internal energy. The power of the energy wave emanating from the region of the core is equal to 10-13 TW. [V.N. Ustyantsev, 2009].

The energy wave structuring mechanism of the Earth's tectonosphere auto-oscillation system leads to its volumetric expansion and the appearance of periodically-discretely located weakened zones, which are performed by volatile and highly charged elements. Hierarchy of energy waves initiates a process of migration, differentiation of mineralogical associations, i.e. promotes their concentration in attenuated zones of various morphology both laterally and radially. The process of formation of high permeability energy zones, both laterally and radially, is genetically connected with the energy wave generated by the Earth's auto-oscillation system.

The element associations are differentiated as follows: highly chargeable (Ti, Sc, Zr, Nb, Ta, Hf, Y, Th, U), noble (Au, Ag). Highly volatile chalcophile elements (Ge, Ga, As, Se, Cd, Sb, Te, Re, Hg. TI, Bi), to the last group should be added H, CO_2, CO, O_2, H_2S, CH_4 and other volatile and easily fusible elements, which rise through permeable fault zones into the crust, where they are localized in the energy zones. Concentration of highly charged elements favours accelerated transformation of rocks of the weakened zone, i.e. mineral raw material deposits are formed. The specialization of the deposit is determined by a complex of factors. Carbon has the ability to join atoms of different elements to form up to three million compounds of all kinds. The systemic properties of carbon, contribute to the formation of mineralogical associations in the structured by energy waves tectonosphere of the Earth's auto-oscillating system.

"The importance of an element in the life of the Earth is by no means determined by its abundance. For example, carbon in the Earth's crust is only 0.1% (by weight). Despite such a negligible content, it plays an extremely important role in the life of the Earth. All biological processes are directly related to carbon. This element is the basis of life on Earth" [5].

Organic matter initially migrates into the Earth's crust in a diffuse form and only thanks to structuring energy waves of both longitudinal and transverse type, generated by the Earth's auto-oscillation system, conditions favorable for processes of concentration, differentiation of useful component, with manifestation of geochemical zoning, are created in the tectonosphere. Otherwise, the substance would remain in a dispersed state.

The density heterogeneity of the energy divisions of the Earth system tectonosphere reflects the cyclic nature of its structural-matter transformation.

Processes occur in the Earth system space which are explained by the basic laws of physics - J. Newton, Archimedes, A. Einstein and others. These processes,.are.responsible for the emergence of:

1) "The Bolshoi Kanimansur deposit (Central Asia). The dislocations are described by strain tensors. The analysis showed the connection of these (T.M. Zlobin) dislocations with dynamic modes of higher hierarchical ranks. These dislocations influenced deposit structure formation and orebody morphology.

The boundaries of inhomogeneous rock complexes, as well as the depths of intracrustal fluid-magmatic sources, 4-8, 11-19, 15-23 km (Kurils), have been identified.

The manifestation of the underlying tectonic-physical mechanism has proved to be similar to that of Kanimansur, where a repeatedly repeated kinematic process of return action has been recorded at depths of 15-25 km. Three tiered upper crustal chambers with ore-bearing solutions at depths of 10-12, 25-28, and 50-55 km are recorded here.

2) ReuzeymanF.M., a fundamentally new theory of fluid ore formation was developed. It is established that "on the background of temperature reduction rich ore formation occurred under the influence of oscillatory, wave-like change of carbon dioxide concentration, in ore-forming solutions. New

thermobarogeochemical criteria for ore body prediction have been developed. The criteria make it possible to predict the location of the ore body, its size, grade and quality of minerals".

3) Decadecan ore node is located in the Central Kolyma gold-bearing region (Grigorov S.A.), "the formation of which took place under the influence of forward and backward (convective) energy and mass transfer, which caused cyclic differentiation of mineral matter at conjugate levels of ore node, ore field, deposit and ore deposit. Ore objects of the ore-forming system hierarchy are reflected in the structures of geochemical fields".

4) By now an extensive factual material on geology and petrography of Alpinotypic ultrabasites and chromitites is accumulated, which makes us interpret their genesis as inseparably connected with tectonic flows in upper mantle (Saveliev 2008), plastic with respect to enstatite, penetrating into the latter along fractures of chipping and tear-off, which may result in tectonic differentiation with isolation of monomineral "layers" of dunite. Chromospinelides during melting of more fusible pyroxenes are also concentrated in essentially olivine rocks, and during intense flow their regassembly in bodies of predominantly linear shape occurs.

Flow textures are almost ubiquitous in the study of chromites, and aggregate growth is often due to rotation. Ore clusters are always associated with dunites, with volumetric chromite-dunite ratios, the extent of ore formation and chromospinelide content in the ore bodies varying greatly. At the same time, petrographic and geochemical characteristics of ore-bearing and ore-free ultrabasites practically do not differ, deposits are not accompanied by geochemical halos" [5].

The leading factor in ore genesis, is the energy factor.

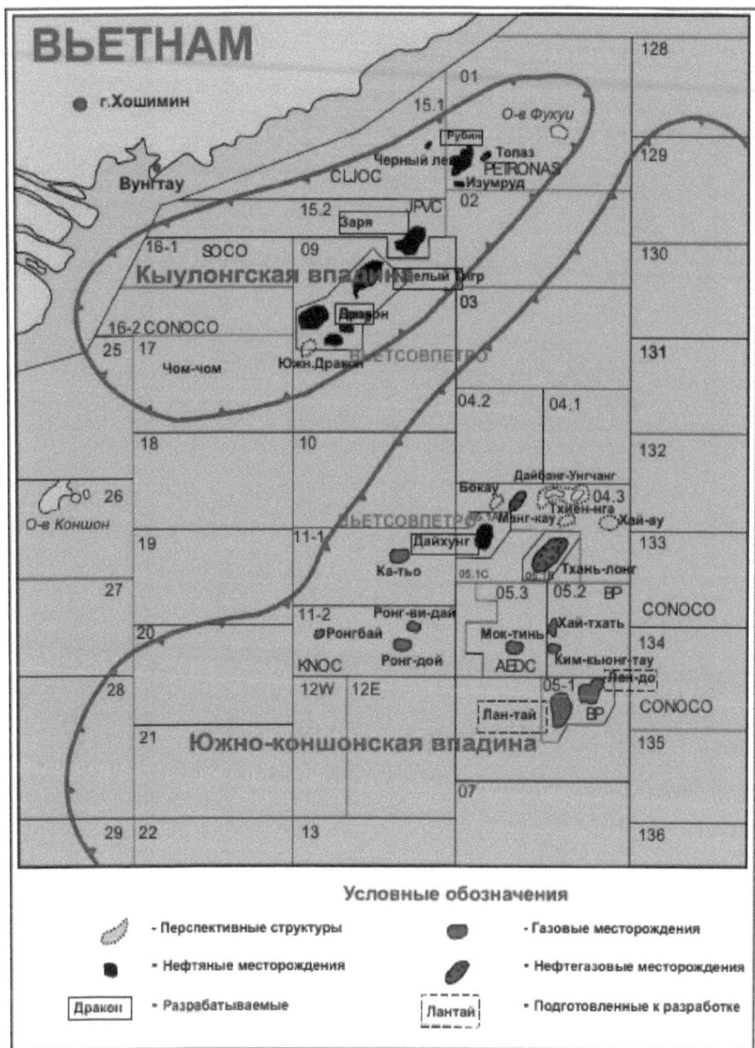

Рис. 1. *Обзорная карта района на шельфе юга Вьетнама*

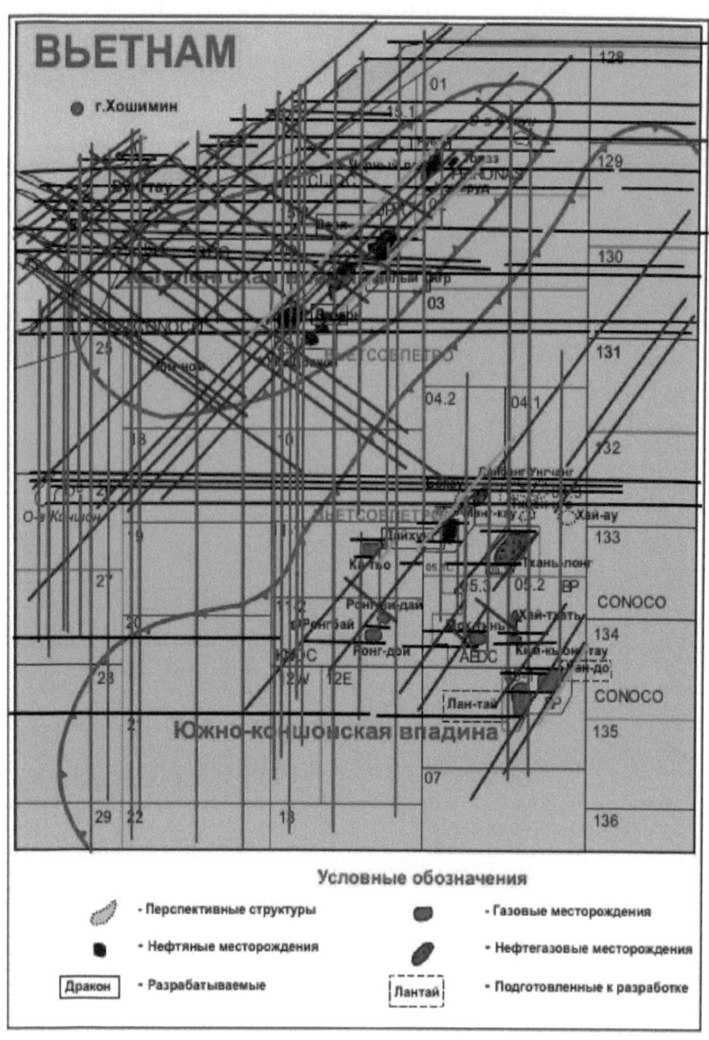

Figure I Overview map of the offshore area in Southern Vietnam. Supplemented by: Forecasting constructs. V.N. Ustyantsev, 2020.

The regular arrangement of the structural elements in the space of the Earth system. The White Tiger oil and gas field. Oil fields are strictly controlled by tectonic faults of northeastern strike, which are characterized by thrusts. The fields are located at the intersection of structures. This fact indicates a genetic connection of oil fields with intrusion rather than dispersed OM in Miocene and Oligocene rocks, (according to O.V. Serebrennikova, 2012) within the limits of the White Tiger field.

One cannot ignore the fact that:

The elemental composition of oil:

C 82.5-87%; H 11.5-14.5%; O 0.05-0.35, rarely up to 0.7%; S 0.001-5.5%, rarely over 8%; N 0.02-1.8%. About 1/3 of all oil produced in the world contains more than 1% S.

Average Corg in the stratigraphic section of the world: Corg=5%, n=50 formations from Paleoproterozoic to Quaternary (V.N. Ustyantsev, 2020).

Compiled by: V.N. Ustyantsev, 2020.

Hydrocarbons are complementary to each other - methane+oil. V.N. Ustyantsev, 2020.

One of Vernadsky's important definitions is that

"the oil accumulated in the fields is a very small fraction of the total mass of oil in non-reservoir formations. To date, there are more than 400,000 known deposits on the planet whose genesis has been attributed to magmatic formations".

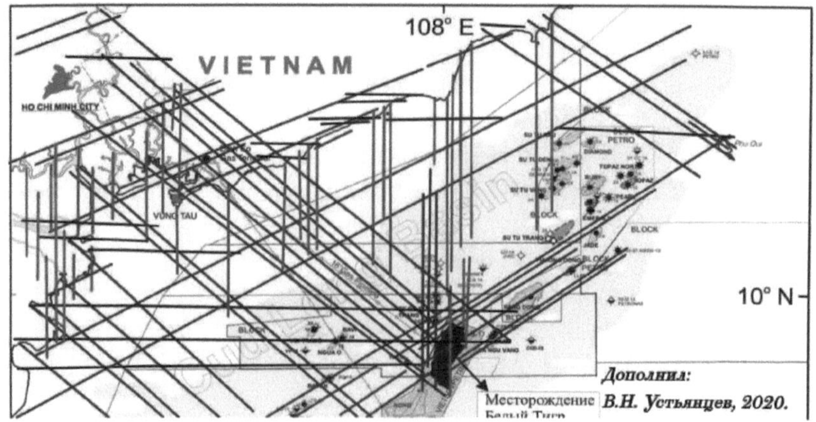

The regular arrangement of structural elements in the Earth system space. The node of intersection of tectonic faults of the four main directions controls the White Tiger field. Compiled by: V.N. Ustyantsev, 2020.

"The system must be open. A closed system, in accordance with the laws of thermodynamics, must eventually reach a state with maximum entropy and stop any evolution" (I. Prigogine). That is to say, the process of mineral formation is anti-entropic. The open system is formed thanks to the hierarchy of tectonic disturbances. So, the zones of tectonic disturbance systems are the main factor under the influence of which the mineral raw material deposits are formed.

The ***particular structure of deep faults*** and their intersection nodes form a closed surface which is an oscillating circuit. The contour is a collector of gases, fluids and magma. Thus, energy-carrying waves entering a heterogeneous medium are reflected and refracted at the media boundaries. These boundaries may cause the appearance of a closed surface from which the waves are reflected which gives the volume, bounded by this surface, vibrational properties and determines the natural periods of the waves, which are typical to the given volume of the vibrational system. In this case the energy of the wave will be given to the transformation of matter. Under closed loop conditions, the velocity of the wave will decrease due to the presence of reflecting surfaces (rectilinear wave propagation is not possible in an inhomogeneous medium).

Along deep faults there are genetically associated weakened resonance tectonic structures - reservoirs of minerals.

The most intense inflow of mantle material is recorded in rifting zones.

Figure A-3. Map of the world basins with estimated shale oil and shale gas formations, as of May 2013 [163].

Regular arrangement of structural elements in the space of the Earth system. Forecasts for hydrocarbons in the North-East of Russia, including the Bering Strait, - the whole region is characterized as an area of ancient and modern riftogenesis and part of the global meridional magnetic East Asian anomaly. Hydrocarbon deposits of the Russian Far East, Japan, Vietnam, and Australia are associated with the riftogenesis zone. Compiled by: V.N. Ustyantsev, 2020.

The establishment of block ***parameters*** determining the localisation of volcanic centres makes it possible to predict the localisation conditions of extrusive, intrusive bodies and deposits associated with volcaniclutonic - complexes.

A generalized idea of the distribution of volcanoes along latitudinal zones also allows us to establish a periodicity with a step of [200] (V.V. Bogatsky), and a similar pattern is outlined in the meridional direction.

In 1968, B.I. Suganov discovered "discrete periodicity in the location of magnetite deposits in southern Middle Siberia".

M.A. Churilin outlines "links of discrete structures with metallogenic and ore fields, nodes, areas, including for intrusions of the central type".

The wave process is well documented in the coal fields. For the central area of the Donbass, V.N. Volkov, established waves with half-wave lengths of 7.6-10; 1.9-2.7; 0.35-0.45 km.

K.V. Gavrilin has observed a dependence for the coal seams of the Kansko-Achinsky basin, where the half-wave is 6-8; 2-4; 0.5-1 km.

"The Earth wave is essentially an enveloping curve that hugs the periodic alternation of maxima and minima in steps of 10o in the zone between 40o N and 40o S. and by 20° step characteristic of higher latitudes. The periodicity of maximums is revealed, connections of which are located in 20o, 40o, 60o .

The similar periodicity in the density of volcanoes (orthogonal network), indicates the same periodicity of the stress fields in the Earth's crust - a 20o step change in the intensity of the stress fields, covering the entire spherical surface of the Earth" [5,7].

Regional seismicity of the space gravitates towards negative morphostructural elements. Hydrocarbon and other mineral deposits are associated with troughs, intermountain troughs, zones of junction of different vault structures, riftogenesis zones.

L.I. Ryazanov pointed out the confinement of oil and gas deposits to structural traps of faults active in the latest tectonic stage (Bukhara-Chardzhou stage, etc.).

M. Valyaev has shown that productive are the intersection nodes of longitudinal and transverse faults, low-amplitude flexure and fault zones and flexures, branches of intracrustal basement faults whose expression up the section gradually fades, but in all cases the faults are characterized by neotectonic and even recent motions.

P.S. Voronov emphasizes that latitudes of 42 and 44 degrees are favourable, especially 43-42 degrees for HC localization.

A number of depth structure features were identified ***by A.A. Borisov 1972***, by recalculating the anomalous magnetic anomaly field at different heights. Recalculations for 100-200 km altitudes indicate a sub-latitudinal anomaly, with positive anomaly fields traced along parallels 70, 56 and 42 degrees and negative anomalies along parallels 65 and 50 degrees.

Research by A. Portnov, 1999

"The kimberlite pipes 'pierce' the thick 40 km crust of the platforms, not the much thinner 10 km crust of the ocean floor or the transition zone - the borders of oceans and continents, where deep fractures contain hundreds of smoking volcanoes and lava pours freely to the surface.

Extensive analytical and experimental material has made it possible to construct a new model for the formation of kimberlite pipes and diamonds. It explains many of the geological mysteries associated with these ultra-deep formations. The model is based on extensive information about the gaseous, predominantly hydrogen-methane 'exhalation' of the mantle, and possibly of the Earth's core.

Kimberlite pipes are traces of 'punctures' in the lithosphere by huge gas bubbles rising from the mantle.

Such a bubble, striving to burst to the Earth's surface, pushes itself thin. "The deep gas pushes them apart under the terrifying pressure of tens of thousands of atmospheres which is transmitted from the mantle to the upper part of the Earth's mantle and then to the upper part of the crust. The deep gas pushes them with the terrible pressure of tens of thousands of atmospheres transferred from the mantle to the upper part of the crust. The kimberlites are confined to platforms, because they are almost gas-tight. Therefore, under them accumulate tiny gas bubbles scattered in the rocks, which merge into large bubbles of hydrogen-methane composition. At a certain critical volume, such a bubble begins to gradually "pop up", i.e. to penetrate into the structure of the platform and to rise to the surface of the planet. Platforms are like saucers floating in an aquarium, with air bubbles rising from the bottom. The bubbles stream around the 'saucer', but some of the gas accumulates under the bottom of the saucer. The gas rises from the mantle, evidenced by the fact that helium here is sharply enriched in the light deep isotope helium. But the underground gases of the platforms contain a thousand times less of this helium than the gases of volcanoes. Consequently, the platforms are a dense blanket for mantle gases. The dispersed mantle gas gathers in big bubbles because of the powerful action of so-called hot spots (geologists have only recently heard about their existence). When big gas bubbles form under the platforms, Archimedes' law comes into play. The gas mixture (hydrogen-methane), even at mantle pressure, would have a density below that of water. But the density of the mantle itself exceeds the density of water by more than three times. It means that the lifting power of the bubble with the volume of 1 cubic kilometre will be 2.5 billion tons! Moreover, this gas is heated to 600-8000 C. The fact that the kimberlite pipes at depth are tapered into a thin stalk tells one that all the enormous lifting force of the gas was applied to a very small area. In the process, dozens of kilometres of rock were pierced like a giant needle. Thus a thin channel 100-150 kilometres long was formed. The gas bubble squeezed upwards until it embedded itself in the soft rocks of the sedimentary cover of the platform. By rising upwards, the gas bubble creates a low-pressure zone at its tail end. Mantle rocks re-crystallised by the gas crush and rush into this zone, with a thin plug. The gas drags the mantle rocks (fluids) behind it. Countless textbooks give diagrams of the diamond-graphite equilibrium and say that diamond arises from graphite. But for some reason no one asked the question: where in the mantle is graphite coming from? After all, it is unstable there and it is called a "forbidden" mineral for the mantle conditions. Carbides are quite a different matter. They are stable here: carbides of iron, phosphorus, silicon, nitrogen, hydrogen.

Hydrogen carbide is a gas, common methane, it is mobile and easily concentrated in the deep fluid. At one time, geologists did not pay much attention to a remarkable discovery of the Soviet physicist B. Deryagin, who back in 1969 synthesized diamond from methane and, very importantly, at a pressure even below atmospheric pressure. This discovery should have by then fundamentally change the existing understanding of the diamond as a mineral which is crystallized necessarily from melts and at high pressures. The data of B. Deryagin allowed me to consider the possibility of diamond crystallization from a fluid, a gas mixture in the C-H-O system.

It turns out that in such a fluid, oxygen at ultrahigh mantle pressure loses its oxidizing properties and does not even oxidize hydrogen. But when the gas rises upwards, when a kimberlite pipe is formed, the pressure drops. It is enough to decrease the pressure 10 times, from 50 to 5 kilobars, for the oxygen activity to increase a million times. And then it instantly combines with hydrogen and methane. Simply put, the gas self-ignites - a fierce fire erupts in the underground pipe.

Example. "Tengiz oil field (Republic of Kazakhstan).

"In 1985, a major accident occurred at the Tengiz field while drilling exploration well No. 37 - from a depth of 4,467 m there was a release of oil and gas into the atmosphere and after a few hours the open fountain of oil and gas caught fire. As a result of the accident, a column of fire, 300 metres high and 50 metres wide, burst from the ground. The temperature around the well was 1,500 degrees centigrade. The fire was extinguished for 398 days, from June 23, 1985 to July 27, 1986. As a result of frequency resonance image processing at the edge of mapped anomalous zone (in the eastern part) detected an anomaly at helium resonance frequencies. Within this anomaly, the estimated reservoir pressure was 113.0 MPa. Taking into account the materials presented above, there are good reasons to consider this local area as a vertical channel of migration of deep fluids". (S.P. Levashov, 2017) [5].

The consequences of such an underground 'fire' depend on the ratio of carbon, hydrogen and oxygen in the fluid. If there is not too much oxygen, it will only vomit hydrogen from the methane molecule (CH_4). The water vapour formed will be absorbed by the mineral dust and form ***serpentinite, the most characteristic mineral of the kimberlites.*** Carbon, left "alone" at a pressure of thousand atmospheres and temperature of about 10000 C will close on itself with unsaturated valence bonds and form a giant molecule of pure carbon, the diamond! In practice, such a favorable combination of components in the gas mixture is rare: only five percent of kimberlite pipes are diamondiferous. More often, there is either too much oxygen or not enough. In the first case, the carbon will burn and turn into gases called oxides: CO or CO_2. In this case, non-metallic kimberlites are formed. They are characterized by increased magnetism, because iron oxide, magnetite, appeared in them. There was a lot of oxygen and it "wrenched" the iron out of the silicates. When oxygen or methane is scarce, only water vapour will emerge and it will be absorbed by serpentinite. It turns out that diamond arises as a product of spontaneous underground combustion of carbonaceous fluid. Methane combustion increases the activity of oxygen and affects the isotopic composition of carbon and nitrogen that make up diamonds, as heavy isotopes are concentrated in the oxidizing environment. Growing diamond crystals trap numerous dust inclusions - tiny grains of minerals from surrounding rocks - from the gas. The age of these mineral inclusions sometimes coincides with the geological age of the kimberlite pipes, but more often the inclusions turn out to be much older. For example, the diamonds from the famous Kimberley pipe (South Africa), which intruded into the surrounding rocks 85 million years ago, have inclusions of garnet-pyrope (determined by the samarium-neariodimium method) that are 3200 million years old. In the Udachnaya pipe from Yakutia, which intruded into the surrounding rocks 425 million years ago, the age of inclusions of the mineral clinopyroxene (determined by the potassium-argon method) is 1149 million years. From such data, geologists usually conclude that diamonds crystallized in the mantle, perhaps billions of years ago, and then an explosion threw them to the Earth's surface. In my opinion, the inclusions in the diamonds were trapped by the growing crystals from the 'dust' of the surrounding gas stream. In recent years, subtle methods of analysis have made it possible to detect among the inclusions in diamonds, native metals such as iron, nickel, chromium, silver and nickel and iron sulphides. How did they get into diamonds?

In my opinion, all of these metals are reduced from the minerals of the underlying rocks - silicates with high iron, nickel, silver and chromium-rich oxides - by powerful reducing agents such as hydrogen and CO, and the underlying hydrogen sulphide has turned some of these metals into sulphides.

The diamond 'armour' preserved this unstable sulphide-metallic dust in the crystals. The sharp 'dry' contacts of kimberlite pipes with the surrounding rocks have long been a mystery to geologists. Geologists know that around magmatic rock massifs of all types there are powerful zones of contact changes due to recrystallization and alteration of the surrounding rocks. But at the contact with kimberlites, changes in sedimentary rocks are negligible. It turns out that there are changes, and very significant ones, but they have an unusual character. Very powerful - up to half a kilometre - halos of concentration of small grains of luminescent minerals appear around the pipes. The content of apatite and zircon grains, minerals luminous in ultra-violet rays, increases tenfold or even hundred times. What is more, apatite glows not with ordinary yellow light, but with bluish light, which is typical of kimberlite apatite. These luminescent halos are explained by strong "blowing" of the surrounding rocks by deep mantle gas with reducing properties and such characteristic elements of kimberlites as europium, cerium, zirconium.

The birth of diamonds, not somewhere in unknown "stone caves" as previously thought, but in the kimberlite pipes themselves, during their formation, explains the almost perfect preservation of diamond crystals, which are found next to kimberlite pebbles, consisting of rounded, chipped and facetless deep minerals actually extracted from the mantle.

The crystallisation of diamonds from gas is also indicated by the continued presence of nitrogen and sometimes boron. There is practically no nitrogen and boron in the mantle silicate melt, but these elements are concentrated in the fluid because they form gaseous compounds with hydrogen. At one time, radon must have accumulated in the fluid. It is radon, the strongest alpha emitter, that may have produced mysterious, unusually beautiful green diamonds. Their colouring is certainly related to the influence of alpha particles. Mantle gas "hung" in the upper layers of the Earth's crust, so no geologists have been lucky enough to find a volcano in the middle of a platform scattering diamond crystals around it. The kimberlite pipes that are found are only uncovered by erosion processes. For the prospector this means that there are many "blind" kimberlite pipes that do not come to the surface. Their presence can be recognized by detected local magnetic anomalies, the upper edge of which is located at a depth of hundreds or, if lucky, tens of meters" [5,7,20].

Conclusion

The principles of P. Curie:

"When certain causes cause certain effects, the elements of symmetry of the causes must show up in the effects they cause". (P. Curie).

"When a certain dissymmetry is found in any phenomenon, this same dissymmetry must also appear in the causes which give rise to it" (P. Curie).

"Positions opposite to these are wrong, at least practically; in other words, the effects may have a higher symmetry than the causes that caused them" (P. Curie).

The primary importance of these provisions, very perfect in all their simplicity, is that the elements of symmetry in question apply to all physical phenomena without exception.

Theorem of I. R. Prigozhin (1947), thermodynamics of non-equilibrium processes:

"under external conditions preventing the system from reaching an equilibrium state, the stationary state of the system corresponds to the minimum production of entropy".

"Synergetics explains the process of self-organisation in complex systems as follows:

The system must be open. A closed system, according to the laws of thermodynamics, must eventually reach a state with maximum entropy and stop any evolution" (I. Prigogine).

"Self-organisation is inextricably linked to wave processes" (I. Prigogine).

"In any open, dissipative and non-linear system, self-oscillatory processes, supported by external sources of energy, inevitably occur, resulting in self-organisation" (Prigogine).

- "When certain causes cause certain effects, the elements of symmetry of the causes must show up in the effects they cause". (P. Curie).

The energy wave, generates a hierarchy of interacting stress fields. The unloading of the stress fields, - deformation, - is accompanied by the generation of a lower order energy wave.

Under the influence of an energy wave, space is divided by a zone of intense degree of deformation into regions of low and high degree of deformation, which also have different energy potentials.

The presence of interacting stress fields in the Earth (space) system space.

"The growth of stress field activity occurs from the poles towards the equator. The ratio of the distance between neighbouring volcanoes to the total linear extent of the volcanic zone is a value reflecting the period of oscillation. This relationship was identified by G. G. Matshinsky in 1951, and a step of 25-32

113

km was defined. The structures of island arcs are characterized by a wave structure".

The works of M.V. Petrovsky, A. Kaye, P. Tricar, have shown that "tectonic structural forms formed in the Earth's crust are displayed in the form of certain landforms. Epeirogenic processes, expressed in periodic deformation of tectonosphere, which are caused by the passage of waves generated in the bowels of the Earth. The oscillations of different orders occurring in the Earth are established by precise instrumental measurements. The summation of the oscillations leads to the phenomenon of resonance".

"The geochemical epochs of formation and localization of minerals are distinguished, and the plains-forming epochs separating them are distinguished" (V.I. Popov) [7].

This global process is a clear confirmation of the cosmogenic factor at work.

Of all known natural phenomena, the system properties of the energy wave are able to structure the Earth system space with the manifestation of patterns of deposit placement in blocks of the Earth's crust. The deposits are located in the blocks, obeying a certain law, that is, the complementarity of the energy wave system properties is shown. As shown in work discreteness, periodicity of placing of deposits of mineral raw materials is shown.

Total power of energy wave coming from the region of core and bottom of lower mantle is about 10 to 13 TW. That is, under influence of energy wave with power from 10 to 13 TW, there is a structural-material transformation of the Earth's self-oscillatory system.

This position is fundamental to understanding the architecture of the Earth system and the mechanism of the processes taking place in its space.

The symmetry of the open, non-linear (auto-oscillatory) Earth system is manifested by the geometrical regularity of the zones of tectonic disturbance systems in the Earth's crust and mantle. Mid-ocean ridges, their geometry, reflect the plan of deformation of the Earth's crust, in space and time - it remains unchanged from the Archean to the Quaternary. This is a geometrical formula showing that there are only four proven strike directions of tectonic structures. This fact allows to successfully apply the method of geometrization of the geoprocess, - the method is very reliable and accurate, as the cosmogenic factor, which is responsible for the laws of location of the objects of space, and thus the structural elements of these objects (location of stationary energy centres, - EEC, faults, deposits), in the Earth system, operates.

The geometric regularity, discreteness, and periodicity of the location of zones of tectonic fault systems, indicates the symmetry of the Earth system. Along the deep faults are located genetically related to them weakened resonance-tectonic structures - reservoirs of minerals.

This circumstance gives a possibility of wide application of analogy method in geology. The method of geometrization of the geoprocess, - is very reliable and accurate, as the cosmogenic factor, which is responsible for the laws of the location of space objects, and therefore the structural elements of these objects (location of ESS, faults, deposits), in the Earth system

- "The system must be open. A closed system, in accordance with the laws of thermodynamics, must eventually reach a state with maximum entropy and stop any evolution" (I. Prigogine).

That is, the process of mineral formation is anti-entropic. An open system is formed by a hierarchy of tectonic faults. Thus, the zones of tectonic fault systems are the main factor under the influence of which mineral deposits are formed.

Elemental composition of oil: C 82.5-87%; H 11.5-14.5%; O 0.05-0.35, rarely up to 0.7%; S 0.001-5.5%, rarely over 8%; N 0.02-1.8%. About 1/3 of all oil produced in the world contains more than 1% S.

Mean Corg in the stratigraphic section of the world: Corg=5%, n=50 formations from Paleoproterozoic to Quaternary analysed. (V.N. Ustyantsev, 2020).

The hydrocarbons are complementary to each other.

With metamorphism, the proportion of C increases and the proportion of H and hetero-elements decreases.

"The principle of 'Onme vivum e vivo' is fully and completely within the framework of the empirical generalisation justified by P. Curic:

- "dissymmetry can only be caused by the same dissymmetry" (P. Curie), writes V.I. Vernadsky. 1934».

Carbon in oil genetically related to the biosphere Corg = 5%.

Petroleum, in the process of self-organisation of the Earth system over geological time, undergoes significant metamorphic transformation - the free hydrocarbons are degraded. The amount of carbon increases and the amount of hydrogen decreases.

This process is largely due to the decomposition of silica-aluminosilicates, which, as shown by Vernadsky (1934), contain large quantities of abiogenous carbon.

"Petroleum, must be considered minerals of biogeochemical, polygenic origin, which have undergone strong metamorphism" (V.N. Ustyantsev, 2020).

"The average carbon content in the Earth's crust, according to A.E. Fersman, is 0.35% (1939). It is determined on the basis of numerous analyses of rocks, natural waters and air. It is clear that such estimates are far from accurate, and the data from different authors differ significantly. However, the order of values of both the total carbon content and its distribution in different zones of the Earth's crust seems to be correct" (V.I. Vernadsky).

The regular arrangement of the structural elements in the space of the Earth system.

Due to the fact that faults are primary structures, they are located linearly and have a through character in relation to other tectonic structures, which makes it possible to successfully apply different geometric methods for prediction purposes.

The paper unambiguously proves the validity of I. Prigozhin's law (1947) and the principles of nonlinear thermodynamics by Prigozhin and the symmetry principle by P. Curie in geology.

The possibility of a sharp increase in the production of the most important ores of rare elements and hydrocarbons lies in the integrated use of mineral raw materials. Knowledge of the regularities of the structure of the Earth's crust blocks and the mechanism of their formation increases the efficiency of geological exploration works and reduces the material costs of their performance, this factor ultimately leads to a reduction in the cost of extracted mineral raw materials.

List of references

1. Vernadsky V.I.. Essays on geochemistry. State scientific-technical mining-geological-oil publishing house. Moscow Leningrad Grozny Novosibirsk 1934.

2. Geology, Geophysics and Development of Oil and Gas Fields. N.L. Kiseleva. Consolidated stratigraphic section of oil and gas-bearing strata of the world. 2017.

3. Subsoil Use in the 21st Century. 2117, No. 3, V.N. Ustyantsev. About geotectomagmatic factor of mineral raw materials generation. Wave Mechanism of Structural and Material Transformations of the Earth System. p. 116.

4. Popov V.I. Mineral Resources of Uzbekistan Academy of Sciences of Uzbekistan Publishing FAN", Tashkent 1976.

5. Ust'yantsev V.N. Energy, degassing of the Earth's auto-oscillation system. On a unified wave mechanism of structure formation and generation of mineral associations in blocks of the Earth's crust. ISBN: 978-5-02-040199-0, Moscow, Nauka Publisher, 2019.

6. Nikolsky B.P. Handbook of the Chemist of the 21st Century, Abiogenic Compounds.

7. Ustyantsev V.N. Origin of primary hydrocarbons and oil. GlobeEdit ISBN: 978-620-0-61141-3.

8.   Akhmedjanov M.A., Borisov O.M. Tectonics to Mesozoic formations of middle and southern Tien Shan. - T.; Fan, 1977.

9.   Belousov V.V., Fundamentals of Geotectonics. - Moscow; Nedra, 1975.

10.   Bogatsky V.V. Mechanism of formation of ore field structures. - Moscow; Nedra, 1986.

11.   Kreiter V.M. Structures of ore fields and deposits. - Moscow; Gosgeoltekhizdat, 1956.

12.   Lukyanov A.V. Problems of physics of tectonic processes. - Moscow; Nauka, 1985.

13.   Yakubov D.Kh., Akhmedjanov M.A., Borisov O.M. Regional faults of Middle and Southern Tien-Shan. - T.; "Fan", 1976.

14.   Rezvoi D.P. Tectonics of the eastern part of the Turkestan-Alay system. Lvov, Publishing house of Lvov University, 1959.

15.   Pokrovskiy A.V. On marginal deep faulting of South-West Gissar. (South Tien-Shan), "Uzbek Geol. Zhl.", 1963, No. 6.

16.   Skaryatin V.D. On studying the fracture tectonics by the complex of various space images of the Earth (method of multistage generalization), Izd. vysokikh uchebnoi zavod, Geol. i razit. 1973, № 7.

17.   Suvorov A.I. Main types of major faults in Kazakhstan and Central Asia. In the book "Depth faults", M., Ndra, 1964.

18.   Suvorov A.I. Regularities of the structure and formation of deep faults. Proceedings of the State Geological Institute, vol. 179, Moscow, Nauka, 1968.

19.   Suvorov A.I. Deep faults of platforms and geosynclines. Moscow, Nedra, 1973.

20.   Science and Life, No. 1, January, 1999, A. Portnov. Diamonds are the "soot" of the mantle.

21.   Petrov O.V. Tectonic map of the Arctic. St. Petersburg, VSEGEI, 219.

22.   Ustyantsev V.N. Fractures in the Earth's crust and the formation of mineral deposits. LAP LAMBERT Academic Publishing, ISBN 978-620-2-923-03-3, 2020.

Ustyantsev Valery Nikolayevich.

Graduated with honours from the Geology Department of TashSU in 1978. Worked in HYDROINGGEO, then in a field geological exploration party (noble metals) (V-K GRE), from 1986 transferred to experimental expedition at SAIGIMS of Tashkent city. He has been engaged in research work since 1974.

I want morebooks!

Buy your books fast and straightforward online - at one of world's fastest growing online book stores! Environmentally sound due to Print-on-Demand technologies.

Buy your books online at
www.morebooks.shop

Kaufen Sie Ihre Bücher schnell und unkompliziert online – auf einer der am schnellsten wachsenden Buchhandelsplattformen weltweit! Dank Print-On-Demand umwelt- und ressourcenschonend produziert.

Bücher schneller online kaufen
www.morebooks.shop

KS OmniScriptum Publishing
Brivibas gatve 197
LV-1039 Riga, Latvia
Telefax: +371 686 204 55

info@omniscriptum.com
www.omniscriptum.com

Printed by Books on Demand GmbH, Norderstedt / Germany